Gabler Theses

In der Schriftenreihe „Gabler Theses" erscheinen ausgewählte, englischsprachige Doktorarbeiten, die an renommierten Hochschulen in Deutschland, Österreich und der Schweiz entstanden sind. Die Arbeiten behandeln aktuelle Themen der Wirtschaftswissenschaften und vermitteln innovative Beiträge für Wissenschaft und Praxis. Informationen zum Einreichungsvorgang und eine Übersicht unserer Publikationsangebote finden Sie hier.

More information about this series at https://link.springer.com/bookseries/16768

Celso Gustavo Stall Sikora

Assembly-Line Balancing under Demand Uncertainty

Celso Gustavo Stall Sikora
Hamburg, Germany

ISSN 2731-3220   ISSN 2731-3239   (electronic)
Gabler Theses
ISBN 978-3-658-36281-2   ISBN 978-3-658-36282-9   (eBook)
https://doi.org/10.1007/978-3-658-36282-9

Responsible Editor: Marija Kojic
This Springer Gabler imprint is published by the registered company Springer Fachmedien
Wiesbaden GmbH part of Springer Nature.
The registered company address is: Abraham-Lincoln-Str. 46, 65189 Wiesbaden, Germany

Acknowledgements

Pursuing a PhD requires a great deal of dedication, energy, and time. Along the path, I found challenges and some successes but also other misfortunes. Although the coding, writing, and setting of the long case studies is mostly a lonely job, several people accompanied me during this four-year path. They deserve my healthful thanks.

To my supervisor, Wolfgang Brüggemann, who heard all ideas and presentations, read and commented on all material: thank you for making a PhD at the Insitute of Operations Research possible. The journey could not be more pleasant. Thanks to the other team members: Sven Pries, for the ideas exchanged and the long discussions, many of them included in the book; Christine Rodenbeck, for all the revisions and all the organization that made working in the Institute so easy; Andreas Geiger, Malte Lübben, Claus Gwiggner, Mehdi Karimi-Nasab, Filip Covic, and Julia Krake for the tips, collaboration, conversations, indications of literature, and company.

To my second supervisor, Malte Fliedner, and the other members of the commission Stefan Voß and Nils Boysen: Thank you for the lessons, the corrections, commentaries, and the interesting questions. To my coauthors and coauthors to be: you made doing research much more pleasant and fruitful.

To my friends in Hamburg, Germany, Brazil, and all over the world: thank you for giving meaning and purpose to such a journey.

To my family, Neide, Celso, Mariana, and Rodrigo: thank you for supporting me from the distance, also when unpredictable events wouldn't allow for visits.

To my fiancée, Dhana: thank you for sharing almost the whole journey with me. Thank you for the love and partnership, that would last for many more chapters of our lives.

Abstract

The automotive industry is highly dependable on assembly lines for the production of today's demand volumes. Assembly lines were once introduced as an efficient production configuration for a single product, in which the production tasks are divided among multiple workstations organized along a conveyor belt. Nowadays, the automotive manufacturers cannot rely on production systems for a single model: the choice of vehicle comes with innumerous configurations, options, and add-ins. In the production site, these different vehicles must share the same resources and may flow on the same assembly line. As a result, assembly lines must be at the same time specialized to provide high efficiency, but also flexible to allow the mass customization of the vehicles.

In this thesis, a compendium of problems and solution algorithms for the assembly line balancing problem considering demand uncertainty is presented. As planning and building an assembly line is a commitment of several months or even years, it is understandable that the demand will fluctuate during the lifetime of an assembly line. New products are developed, others are removed from the market, and the decision of the final customer plays a role on the immediate demand. In this work, the demand or production sequence is modeled using three different view points of a system configuration.

A first approach proposed in this thesis considers total control of the production sequence. In this first problem, the assembly line planner can optimize the assembly line and the production sequence simultaneously. The uncertainty is due to the different time frames of both problems. The planning of an assembly line is a long term decision, while the sequencing problem is solved in short-time based on the customer orders. An exact solution procedure is proposed in this thesis for the optimal design of a paced assembly line, which must operate with uncertain demand to be sequenced in the future. The expected amount of utility work for the production is minimized using a combinatorial version of the Benders' decomposition.

A second problem dealt with in the thesis is the design of an assembly line when the planner plays no role in the production sequence. In this approach, the production sequence is considered to be random. A Branch-and-Bound Algorithm using Markov chains to evaluate partial solutions is proposed and used to solve instances exactly.

A third contribution considers a restriction on the sequence control. The planner has at disposal a buffer to alter the production sequence locally. For this problem setting, the buffer operation is optimized, in which selection policies are proposed and tested. The uncertainty is modeled through a random buffer entry, that must be resequenced respecting production and due date restrictions in an online setting.

Contents

Acronyms

ACEA	European Automotive Manufacturers Association
AGV	Automated Guided Vehicle
ALB(P)	Assembly Line Balancing (Problem)
AS/RS	Automated Storage and Retrieval System
BA(P)	Buffer Allocation (Problem)
BAS	Blocking After Station
BBS	Blocking Before Station
CMS	Cellular Manufacturing Systems
CT	Cycle Time
EU-15	The first 15 countries of the European Union
FiFo	First-in-First-out
FMS	Flexible Manufacturing Systems
GALB(P)	General Assembly Line Balancing (Problem)
MP	Master Problem
MPS	Minimal Part Set
NP	Non-deterministic Polynomial
OICA	International Organization of Motor Vehicle Manufacturers
SALB(P)	Simple Assembly Line Balancing (Problem)
SP	Sub Problem
WIP	Work-In-Process

List of Figures

List of Tables

Introduction

1

1.1 Motivation and Overview

There are few products nowadays that can be compared to an automotive vehicle. Usually weighing from 800 kg to several tons and costing several thousand or even extremes such as a couple of million Euros, it is astonishing to see such a large product playing a major role in our society. According to the European Automotive Manufacturers Association [ACEA, 2020], there were 610 vehicles for every 1000 inhabitants in the European Union in 2018. Such numbers require an annual production of 92.8 million vehicles worldwide or 18.5 million vehicles in the European Union [ACEA, 2020], which is equivalent to almost 3 vehicles per second worldwide. Such production levels require large facilities and a significant part of the labor workforce. In the European Union, direct and indirect jobs in this industry account for 6.7% of the total job market.

Although automotive vehicles and large-scale production existed before, the mass production shift of durable automotive vehicles is credited to Henry Ford with his Ford Model T in 1908 [Binder and Rae, 2020]. The innovation was to consider the transport of the products or workpieces in conveyor belts, on which the vehicles flow through a series of workstations [Binder and Rae, 2020]. The hundreds or thousands of individual tasks are divided among the workers in an assembly line. This way, each worker performs simple tasks in which he or she can specialize. Each worker can then perform the operations within a small cycle time, after which the workpiece is transported to the next station.

The division of the task elements among the multiple workstations is a classical problem in Operations Research named *Assembly Line Balancing Problem* (ALBP) and was firstly discussed in a thesis by Bryton [1954] and in a research paper by Salveson [1955]. A related optimization problem is the *Bin Packing Problem*, in

© The Author(s), under exclusive license to Springer Fachmedien Wiesbaden GmbH, part of Springer Nature 2022
C. G. Stall Sikora, *Assembly-Line Balancing under Demand Uncertainty,*
Gabler Theses, https://doi.org/10.1007/978-3-658-36282-9_1

which objects have to be assigned and fitted into bins in a way that the number of required bins is minimized. In assembly lines, the objects can be seen as the operational tasks, while the bins are the workstations. Instead of having a physical dimension, each task requires a given amount of time in the station. The limitation is not the bin size, but the cycle time of the assembly line. A difference between the Assembly Line Balancing Problem and the Bin Packing Problem is due to precedence relations [Wee and Magazine, 1982]. In the basic version of the Bin Packing Problem, each object can be assigned freely among the bins. A production sequence, on the other hand, usually requires some partial order between the tasks. As products and pieces are assembled, interior parts are not reachable anymore. Hence, there exist precedence relations between the tasks.

The Assembly Line Balancing Problem in its base form as described in the last paragraph is called *Simple Assembly Line Balancing Problem* (SALBP) [Scholl and Becker, 2006] and has very strict assumptions such as deterministic and known production times, a serial line, and the production of a single product. Although these assumptions may be true for Ford's Model T, the automotive market requires a high level of customization nowadays [Boysen et al., 2008]. It is not possible to establish an assembly line for a single- vehicle anymore. Instead, the production system has to be flexible enough to assemble several vehicle variations. Boysen et al. [2008] describe the new paradigm as mass customization, in which the customer can select almost every element of the product from a given range of options. The number of theoretically possible combinations resulting in unique products is huge. Boysen et al. [2009a] report the number of variations of popular vehicle models in 2004, which vary from 40,000 to $3.35 \cdot 10^{24}$ for a selection of European cars.

The presence of multiple product models results in a more complex optimization problem, since not only the assignments of tasks to stations are important, but also how the product models are sequenced. Different production layouts, the presence of buffers, and the production sequence greatly affect the productivity of an assembly system. Furthermore, the customer's taste changes and evolves. So the demand itself may vary during the operational time of an assembly line. Such complexity factors are explored in this manuscript, mainly dealing with the uncertainty of demand in the balancing of multiple-product assembly lines.

1.2   Objectives and Document Outline

The outline of the document is described along with the objective of each chapter. In general, the thesis brings new contributions to the research of assembly line balancing under demand uncertainty.

The first objective of the document is to describe the production stages at automotive manufacturers. The production of vehicles is different than considering a general product because of the large dimensions. Cars, trucks, or buses are large and heavy products, so their handling is rather limited. The deviations of production times cannot be easily compensated by buffers, since the size of the products poses a strong restriction. In Chapter 2 the different stages of production are described, among related optimization problems, such as production planning, assembly line balancing, sequencing, resequencing, etc.

Chapter 3 contains a literature review on different sources of uncertainty in the balancing of assembly lines. A classification of the literature is extended to model the stochastic components of the problem. The uncertainty is mostly modeled in the processing times, while much fewer references deal with uncertain demand or production sequences. Chapter 3 is also used to identify gaps in the literature, which are partially filled by contributions described in Chapters 4–6.

The research core of the manuscript consists of three chapters containing each a problem definition and a solution procedure. All of the contributions deal with assembly lines under uncertain demand, although in each chapter a different assumption or view of the problem is proposed. One key aspect to distinguish the three problems is the control over the production sequence.

The first contribution is detailed in Chapter 4. For this problem setting, production sequencing is totally defined by the planner of the assembly line. This assumption allows selecting a production sequence that matches well with the assignment of tasks in the assembly system. For this problem, the assembly line problem and production sequencing problem are solved in an integrated form. As both decisions are taken in different time frames in practical applications, a hierarchical approach is defined. The assignment of tasks to stations is a medium to long-term decision, while the production sequencing is solved on a daily or weekly basis. The uncertainty in the problem is represented by an uncertain demand at the planning stage of the assembly line. This way, the balancing of the assembly line has to be defined before the realization of the demand, while the sequencing can be solved after the customers define their orders. The problem is defined in a two-stage stochastic programming model, for which an exact solution procedure is proposed to minimize the expected utility work (amount of work from auxiliary versatile workers). A Benders' Decomposition Algorithm based on combinatorial cuts is developed among valid inequalities and improvements. The contents of Chapter 4 has some overlap to the published article version of the chapter (see Sikora [2021]). The results of both publications, however, are complementary.

At the other end of the control spectrum, the second approach models the balancing problem under no control over the sequence. In Chapter 5, the production sequence is considered to be random under given probabilities of the customer purchasing each individual option. It is shown that the stochastic evaluation of a balancing configuration can be modeled as a Markov process, yielding the exact computation of the expected utility work under a random sequence. A Branch-and-Bound Algorithm is used to solve the problem, for which not only the assignment but also the station length is optimized. The objective function is modeled as the sum of the cost contributions of the expected utility work cost and the line length cost. The solution procedure consists of a three-stage process: in the first stage the assignments are defined; the second stage is a single variable optimization for the station length; while the last one is a Markov process to calculate the expected utility work for a given assignment and station length.

In between the sequence control paradigms, a third work examines the implementation of resequencing in automotive assembly lines. In Chapter 6, the problem of restricted resequencing under an uncertain production sequence is tackled. The scope of the problem is modeled within a buffer at the beginning of the assembly line. For a given pre-determined assembly line, the problem consists of selecting the production order to enter the production system. The considered production cost contribution is the utility work necessary for the production within the cycle time. In each cycle time, a product in the buffer must be selected, while the next product entering the buffer is considered to be random. The product selection is performed for every cycle time without the knowledge of the new entering vehicle model. The described situation is an online decision problem, for which heuristic policies are proposed and tested using a simulation model.

The thesis ends with a conclusion in Chapter 7. At the conclusion, the results of the chapters are summarized among with an evaluation of the fulfillment of the proposed research objectives and future works.

Design and Operation of Assembly Lines   2

In this chapter, assembly lines are introduced and described comparing their features to other production layout organizations. The contextual characteristics of the automotive industry are described, along with their repercussions on the design and operation of assembly lines.

2.1   Production and Layout Configurations

Production is a process in which an object is transformed in order to increase its value [Günther and Tempelmeier, 2012, p. 7–8]. This transformation has raw materials or semi-finished products as inputs and finished goods as output. The production of screws, cell phones, cars, and airplanes, for instance, starts with the extraction of ore and oil, which are transformed into several intermediary products until the final assembly of the product. Most of the goods produced in the current economy are made by industrial production divided into several levels, industries, and companies [Günther and Tempelmeier, 2012, p. 2]. In the era of globalization, the production steps may be performed even on different continents. For the production of a car, for instance, it is not unrealistic to think of ore extracted in Brazil being processed in steel sheets in China, transformed into parts of a transmission in Germany, which is finally assembled into a car in Mexico for its local market.

The process of a given production step can be classified based on the number of different products and the produced quantities. Günther and Tempelmeier, 2012, p. 11–12 characterize the production as mass, serial, and individual. The production of a single good (or a family of similar goods) continuously and indefinitely is characterized by mass production. The homogeneity of the products is ideal for a highly efficient production often mechanized or automated. Examples are the production of cement or screws, which are industrially produced by dedicated resources. Serial

C. G. Stall Sikora, *Assembly-Line Balancing under Demand Uncertainty,* Gabler Theses, https://doi.org/10.1007/978-3-658-36282-9_2

production is a process within mass production characterized by the production of batches of given sizes. That is, the resources of a production system are devoted to a good for a specific amount of time in which the batch is produced. A set-up is then used to adapt the production system to a new product, initiating a new batch. Serial production often uses highly specialized machinery and workers but may be flexible enough to adapt itself to different products. Finally, individual production is responsible for non-standard products that may be specified differently in each customer order.

The production process differs not only between industries but also within companies themselves. An efficient process usually requires different machines and specialized workers, which are allocated to multiple steps in the production process. Suitable production systems depend on the quantity and kind of goods produced. On the one hand, an airplane may take months to build and is usually highly customizable. On the other range of the spectrum, screws are standardized goods that are produced in a number of millions per day. The complexity, demand volume, and customization options play an important role in the adequate and profitable way to produce these goods.

Among the most important production configurations are the function-oriented and the product-oriented layouts. In a function-oriented configuration, the resources are divided based on their function, in which similar machines are spatially set up together. For instance, turning, drilling, and milling operations are performed in separate spaces using the respective resources. This configuration allows the production of a high variety of goods since products can be moved to the necessary machines in the required order. This degree of flexibility, however, imposes limitations on production levels (in comparison to the product-oriented configuration, presented next). The complex flow and the non-specialization of the production resources are therefore more adequate to the production of small or single-product batches.

At the other end of the spectrum, product-oriented layouts are based on the produced good. Machines and resources are organized in the order in which they are necessary for the given product. This organization simplifies the material flow since each step of the production is performed by a machine or station along with the production site. This layout is ideal for highly specialized and automated machinery and is applicable when all production steps are based on a single product or a family of similar products. The productivity of the desired product is high but its flexibility is rather limited. Therefore, product-oriented layouts are more adequate for the mass production of a single good or the production of large batches.

There exists also intermediary configurations such as production cells, manufacturing centers, and flexible production lines. These configurations share partly the characteristics of the two layouts. One schematic representation of the adequate layout based on the number of different goods produced (variants) and the yearly volume of production is illustrated by [Arnold et al., 2008, p. 124]. There is a trade-off between flexibility and productivity since the product-oriented layouts implemented in transfer lines are very efficient but rigid, while pure function-oriented layouts as in a workshop can produce a high amount of different products, but are limited in their efficiency.

2.2   Production and Layout Configuration in the Automotive Industry

The automotive industry represents a considerable part of the world economy. Just in the EU-15, the first 15 countries of the European Union (until April 2004), 428 billion Euros are collected in taxes due to motorized vehicles [ACEA, 2019]. In 2019, 326 million cars are estimated to exist in Europe [ACEA, 2019], while according to the International Organization of Motor Vehicle Manufacturers [OICA, 2018], in 2018 the yearly production of motor vehicles exceeded 95 million. The two leading car manufacturers in production quantities, Toyota and Volkswagen, produced each over 10 million vehicles in 2017 [OICA, 2017]. These production numbers help to justify the selection of production layouts of the larger automotive manufacturers described in this section.

Nowadays, the completed production of one vehicle per minute is typical for the automotive industry [Emde and Gendreau, 2017]. At the same time, there is market pressure for variety and customization possibilities [Boysen et al., 2009a]. The number of configurations, colors, features, and add-ins that are available can be combined to theoretically trillions of different vehicles [Boysen et al., 2007]. The large production level under high varieties is also described as mass customization [Boysen et al., 2008].

Although some luxurious brands still rely on individualized and mostly manual production, the majority of the vehicles are produced in a highly specialized product-oriented layout. The production phases of a vehicle performed by a vehicle manufacturer are mainly divided into 5 steps: Press shop, Body-in-white, Paint shop, Power train, and Final assembly [Omar, 2011, p. 1–14]. It is noteworthy that only a subset of the production operations is performed by the automotive manufacturers [Meyr, 2004]. Most of the components and parts of the vehicles are delivered by suppliers and only assembled by the manufacturer.

A simple illustration of automotive production is given in Figure 2.1. The initial processing steps are performed in the press shop, where sheets of metal are molded to form the pieces of the structure of the vehicle. This part of the process depends on heavy machinery and is highly automated. The Body-in-white part of the production is responsible to join and weld together the output pieces of the press shop. A car may require from 3,000 to 7,000 spot weld points to hold its final form, depending on its size [Hamidinejad et al., 2012; Sikora, Lopes, Schibelbain and Magatão, 2017]. The welding procedures of the body-in-white are also highly automated [Michalos et al., 2010] with multiple robots operating in several workstations [Lopes et al., 2017; Michels et al., 2018].

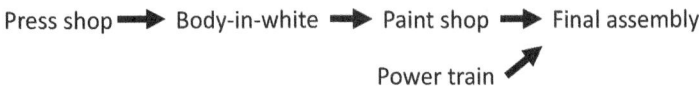

Figure 2.1 Categorization of the processes of automotive manufacturers

The next step of the process is the painting of the body-in-white. This process is critical to the sequence of produced vehicles, since changing the color of the paint may require a set-up time to clean the tools [Boysen et al., 2009c]. An example of a sequencing optimization regarding the paint shop can be found in Gagne et al. [2006]. The Power train assembly is a pre-assembly process that can occur in parallel to the body-in-white and paint shop production. The power train consists of the engine and the drive train of the vehicle. The power train and the body-in-white are then assembled in the final assembly.

The last step of the process is the final assembly, in which most of the components are mounted in the vehicle. Among the assembled elements are all the interior parts, lights, wires, tires, windscreens, dashboards, etc. Several of the assembly procedures are still manually performed due to the great variety of tasks and products. The assembly line balancing problem, the focus of this work, is mainly based on this last step of the production.

To achieve production rates of one vehicle in less than one minute, the necessary tasks must be divided into several workstations. Assembly lines may have hundreds of workers to ensure that the aimed production rate is achieved. The rapid production speed requires a huge amount of logistic and organizational efforts. The employed solution for the transport flow between workstations is mainly realized by conveyor belts. The transportation can be implemented in a paced or an unpaced system. Paced assembly lines have a conveyor belt running at a constant speed. The selected speed controls the cycle time, that is, the interval time between the production of two vehicles. The workstations are organized in the physical space along the assembly line. The workers perform their required operations in the product that is located at his or her stretch of the conveyor belt at each time. After finishing a product, the worker can start the operations of the next product that enters the workstation.

One alternative to the paced conveyor belts is the implementation of unpaced systems. The unpaced transportation of pieces does not rely on a constant speed of a conveyor belt. In these systems, pieces remain in the workstation until it is free to move to the next station. Therefore, the movement is said to be unpaced. There are two categories of unpaced systems, the synchronous and the asynchronous lines. In synchronous lines, all products located at the line are moved at once, after all tasks of each workstation are completed. An asynchronous system allows products to be moved independently. The product can be sent to the next station as soon as the operations are finished and the next station is free to receive a new product [Lopes, Michels, Sikora, Molina and Magatão, 2018].

The organization of factories on flow-systems such as assembly lines poses hard constraints on the sequence in which products can be produced. In simple conveyor belts, the products are moved from station to station in the same order. Rearranging the sequence of products as large as cars or trucks requires extra handling systems that may not be available at each station or assembly line. Furthermore, the size of the products of the automotive industry poses limitations on the use of buffers in the production. Although some buffers are usually used between the paint-shop and final assembly (see Section 2.6), buffers are rarely used within an assembly line due to their costs and size. The low availability for buffers makes the system strongly reliable on the synchrony of the workstations: the output of a station is the direct input of the following station. Because of this, failures or errors in a single station may cause the interruption of the production in the whole assembly line.

To cope with the challenges of high production levels, high variety of products, and high reliability, the automotive manufacturers rely on the even division of the workload among workers who are highly specialized in a small set of tasks. Further-more, large importance is given to the supply of the workpieces and tools at each station. Since a huge variety of products can be produced, manufacturers usually sort the necessary parts in the production order and use the logistic system to deliver these when needed. According to Boysen et al. [2015], the materials are organized in kits to ensure easy access without affecting the processing time. The logistic of the part supplies is classified as Just-in-Time since the parts should arrive when they are needed, or Just-In-Sequence, that is, the pieces are delivered in the sequence they are needed [Boysen et al., 2009c]. These conditions require a robust supply chain coordination between the manufacturer and the suppliers.

2.3   Balancing of Assembly Lines

As already mentioned in the description of the production of automotive manufacturers, the division of the workload among the production resources is essential to achieve high production rates. The choice of a division of processing activities or tasks among resources or workstations in an assembly line is called "assembly line balancing". Not only the division of tasks is important in the industry, but also the formulation and solution of the assembly line design as an optimization problem is well explored in the research community. The first work that appeared in a research journal occurred in 1955, published by Salveson [1955].

In the literature, the optimization problem of the division of tasks in assembly lines is named *Assembly Line Balancing Problem* (ALBP). As its name already clarifies, the main objective is to balance the tasks among multiple workstations to smoothen one or more criteria. Mostly, the total processing time in each station is the aim of the balancing. There are, however, other possibilities, such as the balancing of ergonomic efforts among workers [Otto and Scholl, 2011], space needed for equipment and tools [Bautista et al., 2016], failure risk among machines [Müller et al., 2018], or variability in the case of stochastic processing times [Kao, 1976].

Among various characteristics a task can have the most relevant to most balancing problems are the processing time and the precedence relations [Baybars, 1986]. The processing time is the amount of time a task needs to be performed. Although the processing times may vary from worker to worker, the highly specialized task assignments in the automobile industry usually result in stable and deterministic processing times [Falkenauer, 2005]. The second element is the precedence relations between pairs of tasks. In the assembly process, physical or technological restrictions may require that tasks are performed in a specific order [Gutjahr and Nemhauser, 1964]. Such restrictions are named precedence relations, which can be drawn as a graph. An example of the data for a balancing problem with 7 tasks and precedence relations is given in Figure 2.2. The number inside each circle represents the task number, the processing time is represented at the upper-right corner of each task, and the arrows between circles represent the precedence relations. Directly linked tasks, such as tasks 1 and 3, define direct predecessors (task 1) and successors (task 3). The precedence relations are also transitive, so that indirect precedence relations, such as tasks 1 and 5, also apply. In this case, task 1 is called an indirect predecessor of task 5. In the assembly line, a precedence relation means that a predecessor task must be performed before its successor. As the stations are sequenced along with the handling system, predecessors must be performed earlier than or at the same station as its successors.

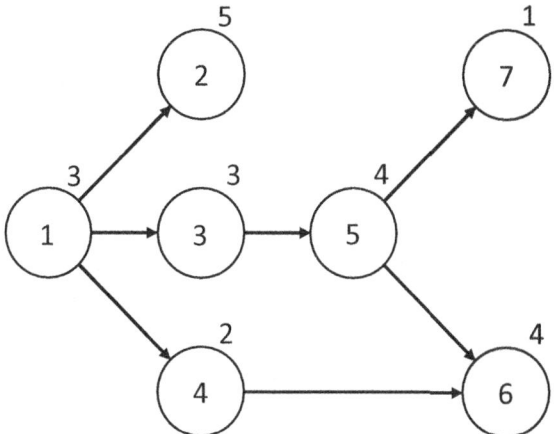

Figure 2.2 Example of a precedence diagram

The balancing problems are classified based on the extension of their characteristics. Baybars [1986] distinguishes the so-called "Simple Assembly Line Balancing Problems (SALBP)" from the "General Assembly Line Balancing Problems
(GALBP)". The SALBP version of the problem is based on various simplifications
and assumptions and its applicability on real assembly lines is limited. However,
several optimization algorithms rely on simplifications of the SALBP's structure, so
that the development of models for the simple version of the problem is still relevant
[Scholl and Becker, 2006; Battaïa and Dolgui,2013].

Baybars [1986] enumerates eight assumptions for the definition of SALB problems:

1. Processing times are known, static, and deterministic
2. Each task must be assigned to a single station
3. There may be precedence relations between tasks
4. All tasks must be performed
5. All stations are equally equipped and manned. No restriction such as station
   height, position, or worker ability applies
6. The processing time of a task does not depend on the station on which it is
   processed
7. The line is assumed to be serial and to have no feeder lines (subassembly parallel
   to the main line)

8. A single product model is produced.

One important consequence of these assumptions is that the production rate of the assembly line is easily determined. As the processing times are independent of workers and stations and only one product model is assembled, the processing time of each station is simply the sum of the processing times of its tasks. The most loaded station is the bottleneck of the entire line so that the cycle time of continuous production is limited by the largest station processing time. An example for a feasible solution of a balancing problem is shown in Figure 2.3. A set of eleven tasks are assigned to six workstations. All tasks are assigned to only one station and precedence constraints are observed. The processing time of each station is shown at the top-right edge of each box and is computed by the sum of the processing times of the tasks assigned to the station. Station 3 has the largest processing time (11 time-units) so that this bottleneck station dictates the speed of the whole assembly line. In the other stations, idle time occurs when a worker has finished the operations and waits for the next workpiece.

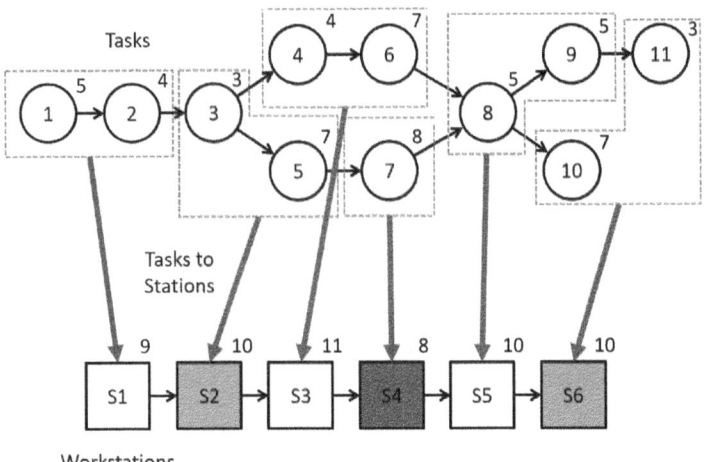

Figure 2.3 Example of a balancing solution

The additive nature of the processing times for the SALBP is important for the definition of the objective functions of the problem. Among multiple variants, the

2.3 Balancing of Assembly Lines

most explored ones are named SALBP-1 (type-1) and SALBP-2 (type-2) [Scholl and Becker, 2006]. In a type-1 problem, a given production throughput is given as a parameter. The objective is to assign the tasks in a way that the number of stations needed is minimized. In other words, the running costs of the assembly line are minimized for a given production level. In SALBP-2, the production level is a variable of the problem. The number of stations is given, while the objective is to maximize the use of the given resources by minimizing the cycle time. Other versions and objectives are also possible, such as the maximization of the line efficiency (SALBP-E) and the feasibility version of the problem (SALBP-F) [Scholl, 1999]. The latter focuses on whether a feasible solution exists for a given cycle time and number of stations.

A General Assembly Line Balancing Problem is defined when one or more of the assumptions of Baybars [1986] are relaxed. There are examples of variations of the problem changing each of the listed assumptions. The first assumption is relaxed when the processing times are considered either dynamic in the case of learning effects [Chakravarty, 1988] or stochastic [Kottas and Lau, 1973]. Tasks that require long processing times may be assigned to more than one station. This may be implemented by using parallel stations that can be realized by, for instance, two workers working the double of the cycle time alternatively in two adjacent stations [Becker and Scholl, 2009]. The third assumption does not necessarily need to hold. An assembly line balancing problem without precedence relations is equivalent to a bin packing problem [Wee and Magazine, 1982]. The fourth assumption requires that all tasks are performed. Although it is generally the case, there are also variants in which only one option between a set of tasks may be performed. These options are called alternative subgraphs and are dealt with in Scholl et al. [2009]. The stations may be differently equipped or manned in real assembly lines, disregarding assumptions 5 and 6. Versions of the ALBP may employ equipment with different costs and capabilities to be chosen [Rubinovitz and Bukchin, 1991] or heterogeneous workers [Miralles et al., 2007]. Other configurations of assembly lines can also be modeled, such as feeder lines, U-shaped lines [Miltenburg and Wijngaard, 1994], parallel lines, lines with moving workers [Sikora, Lopes and Magatão, 2017], etc. Finally, assembly lines for multiple products as in Thomopoulos [1967] are extremely common in real production systems [Boysen et al., 2008].

One important consequence of relaxing some of the assumptions is that the calculation of the cycle time may not be additive anymore. One example of this property is when more than one worker is assigned to a station. Figure 2.4 illustrates multiple possibilities of assigning workers to stations. The simple version of the problem assumes a single worker in each station. Two-sided assembly line problems use the advantage of large products of the automotive industry and allow two workers at the same time [Bartholdi, 1993]. In these lines, tasks are distinguished by whether they can be performed on the left, the right, or both sides of the car. Since there may be interactions between workers, the tasks should be scheduled within one station. These interactions may be due to precedence relations (such as a task performed by one worker is necessary for a successor task performed by the second worker) or accessibility restrictions (when the tasks of both workers require the same physical space to be performed). This necessity of scheduling changes the processing time calculation of the station. Instead of the mere sum of the duration of the assigned tasks, the processing time in the station is the maximal amount any worker needs to finish his or her operation, including possible waiting times. Other variations with more workers are also possible, such as the *Multi-Manned Assembly Line Problem* [Becker and Scholl, 2009; Michels et al., 2019]. In this variation, more than two workers may be assigned to a station simultaneously. Other examples of variations in which the cycle time is not the simple sum of the individual task processing times are assembly lines with set-up times between tasks [Akpinar et al., 2017] or assembly lines with multiple product models. The latter is explored in more detail in the next section.

2.4   Influence of Multiple Products

Designing assembly lines for multiple product models increases the number of features to consider comparing to single-product lines. Cars with different seats or consoles may need more or less time for their installation. Even greater is the impact of the installation of a sunroof: some products may require its installation, while others without it do not need any processing time at all for this task. To be efficient, the balancing of an assembly line with multiple product models has to consider all the possible products and their variations.

In this section, the influence of multiple products in the balancing solution is illustrated. For that, a small instance with three stations and four product models (M1 to M4) is considered. The processing times per station and product model for a given balancing solution are given in Table 2.1. Note that products may have the same processing times in the same station, for instance, product models 2 and 3 in station 1 or 2, or processing times that may greatly differ, such as product models 1 and 2, in station 1. These differences may be justified by a different set of operations assigned to the station or by the different duration of the same tasks (such as the seat or sunroof example). For example, Thomopoulos [1967] assumes that a task must be assigned to a station for all products. These assignments are more restricted than freely assigning tasks differently for each product. This constraint, however, may

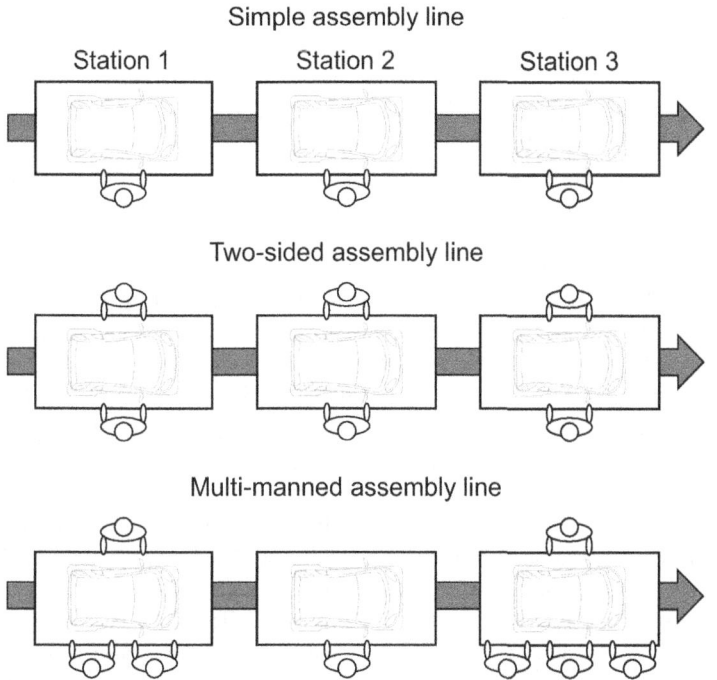

Figure 2.4 Comparison of a simple assembly line, a two-sided assembly line, and a multi-worker assembly line Michels et al. [2019, Fig. 1]

Table 2.1 Data for an assembly line instance with three stations and four product models

Station	Processing time			
	Product 1	Product 2	Product 3	Product 4
1	5	2	2	4
2	4	3	3	2
3	3	2	3	1

bring a large cost reduction. Equipping and training personal of multiple stations for the same task may significantly increase the running costs of an assembly line [Thomopoulos, 1967].

The observed effect of multiple product models depends on the handling system of an assembly line. Therefore, paced and unpaced lines are discussed separately, starting with unpaced asynchronous configurations. In these lines, the movement of a product to the next station is allowed when all tasks are finished in the station and the next station is free. Figure 2.5 is a Gantt chart with the temporal description of the production of the four product models (M1-M4) in ascending order. The products flow through the line in the same order and no overlapping is possible. Each row represents a station, while each gray block represents the processing time of a product. The length of each block is the same as the processing time data given in Table 2.1. At time 0, station 1 starts the processing of M1, which takes 5 time units (t.u.). At time 5 t.u., M1 is finished in the first station and is moved to station 2. Simultaneously, station 1 receives the next product (M2) and starts its operation. As is common in the literature, the transport times are neglected, since they normally just add a constant transfer time to each station (exceptions are found in Bard [1989]; Michels et al. [2018].) Product M2 is finished at station 1 at time 7 t.u. At this moment, however, station 2 is still processing M1. Therefore, M2 cannot be transported to the next station and has to wait until the conclusion of product M1 at time 9 t.u. This condition is named blocking since product M2 is impeded to be moved ahead. Another phenomenon that occurs in unpaced asynchronous lines is called starvation. A station is said to starve when it has already completed the assembly of a product, but the next product is not ready to be transferred yet. Starvation occurs between time 14 t.u. and 15 t.u., while station 3 has finished and passed product M2 along and waits for M3. In this diagram, blocked tasks are shown as white blocks with lower case letters (m_2, m_3, and m_4) while starvation is shown as blank spaces between blocks of tasks.

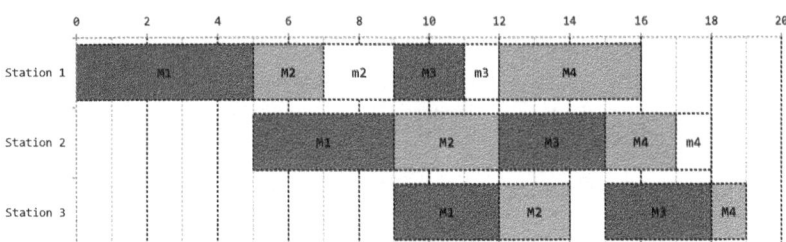

Figure 2.5 Gantt chart of the production in an asynchronous unpaced line

As illustrated in Figure 2.5, the working times of workstations of unpaced assembly lines are affected by the different processing times and the waiting times. The four products occupied station 1 for a total of 16 t.u., while the sum of their processing times would require only 13 t.u. according to Table 2.1. The production rate of such lines is also not constant. Taking the difference in time between the production of each two adjacent products in Figure 2.5 after the time 9 t.u., products take respectively 3 t.u., 2 t.u., 4 t.u., and 1 t.u. to exit the assembly line. Note that, in the Gantt diagram, stations 2 and 3 do not produce until M1 arrives at these stations. This only occurs because of the example simplification: the line is considered initially empty. In practice, much more than 4 products would be produced and all stations are likely to be occupied almost at any given time.

In unpaced systems, the start, end, and move times of each piece depend on the combination of the processing times of all products. The production rate is variable and is also strongly affected by the sequence of products. The system is, however, flexible to deal with varying processing times, since the movement of pieces can be done independently.

A second possible handling system is a synchronous unpaced line. Synchronous lines are similar to the asynchronous unpaced line of Figure 2.5, but they present a further movement restriction. The handling of pieces is done simultaneously. That is, all pieces are moved to the next station at the same time when all of them are finished. A Gantt diagram for the example in a synchronous line is shown in Figure 2.6. The transfers of pieces are done at 5 t.u., 9 t.u., 12 t.u., 16 t.u., 19 t.u., and 20 t.u. Note that even though products M2 and M3 are finished in stations 3 and 2 by time 14 t.u. and 15 t.u., respectively, the synchronous movement requires that product M4 must be finished before the transfer of all pieces. Synchronous lines have no starvation, but instead, all non-bottleneck stations are blocked until the synchronous transfer.

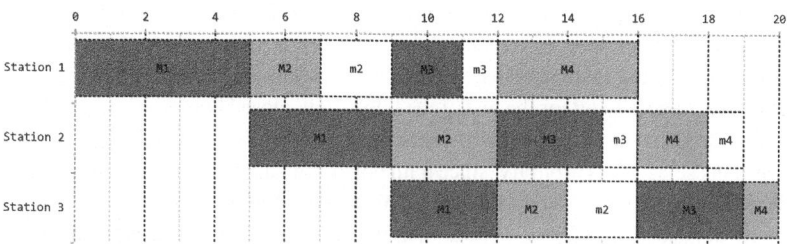

Figure 2.6 Gantt chart of the production in a synchronous unpaced line

Just as asynchronous lines, the productivity of synchronous unpaced lines is variable. Due to the extra restriction of simultaneous movement, synchronous lines are less than or in the best case as efficient as asynchronous lines. Hybrid variants can also be considered, in which part of the line moves synchronously and part asynchronously with intermediate efficiency [Lopes, Michels, Sikora, Molina and Magatão, 2018].

Paced lines operate with a conveyor belt that continuously moves the products along the line. Workers can perform the assembly tasks during the time in which the product flows within the workstations. The operation of a paced line is defined based on three parameters: conveyor belt speed, launching rate, and station length. The conveyor speed controls how fast the product flows along with the workstations and is normally constant along the assembly line. The launching rate is defined based on the time interval between the entry of two products in the line. This time interval is also defined as the *cycle time* Boysen et al. [2007]. The launching rate is usually constant [Boysen et al., 2009c], although they can also vary [Fattahi and Salehi, 2009]. Finally, the station length is the physical space defined for a workstation and can be defined as a time-window in which the workers can access the product [Yano and Rachamadugu, 1991]. Note that the three parameters are interconnected: a given cycle time and conveyor speed may define a minimal station length, for instance. If the station is shorter than the multiplication of the cycle time and the conveyor speed, the worker's window is shorter than the cycle time. He or she may perform tasks during the period in which the product is available and may wait for the next product at the next launching time. For a given length, station's limits can be defined, which represent the initial points and the final points between which workers may perform tasks for a product. These borders between stations can be further distinguished between open and closed borders. The latter present fixed limits on the space of a station. Open borders are soft restrictions, that is, the worker may extrapolate the station limits and advance into the adjacent workstation. Working outside of the window, however, may cause interference with workers of the adjacent stations [Yano and Rachamadugu, 1991]. In the example with the processing time data from Table 2.1 and for the rest of the book, only closed borders are considered. Furthermore, a conveyor speed of 1 length unit (l.u.) per time unit (t.u.) is assumed here. This selection is arbitrary, since a line operating at double the speed and with the double length has exactly the same time window. In a real setting, the station length and speed are set to a value so that the equipment and tools fit the available space [Chica et al., 2013].

Operating paced lines with closed borders and a constant launching rate means that the amount of time each product remains in each station is equal. This poses some difficulties to the balancing of multiple product models because the necessary processing times vary with the product. One of the alternatives would be to select a cycle time so that every product could be completed in the worker's time window. For the data of Table 2.1, a cycle time of 5 t.u. and a station length of 5 l.u. (conveyor belt speed is set at 1 l.u. per t.u.) would be required so that M1 can be finished in time at the first station. Although this solution is very robust, since it is based on the worst possible case, it is very inefficient. M1 has idle times in the other two stations while the processing of products M2 to M4 causes idle times in every assembly step. Product model M4 in station 3, for instance, is processed only in 20% of the time of a cycle.

A second alternative is to select a cycle time and station length based on the average needed for production in the bottleneck station. For the example of Table 2.1, the first station requires an average of 3.25 t.u. per product. Setting this launching rate is roughly 54% more productive than the worst-case solution. This higher production, however, can not always be realized. In the first station, M1 and M4 require more time than the time available in a cycle (the same happens to M1 in the second station). That is, the production may be infeasible without any additional measures to compensate the processing times longer than the cycle time.

A combination of a longer station length and remedial actions can provide the flexibility needed to compensate for the product model variations. Figure 2.7 illustrates how the length of the station can be used as a time buffer for the production of the example of Table 2.1. In the figure, the position of the worker on station 1 is displayed considering a station length of 5 l.u.. The worker starts in the first row by the processing of M1, which takes 5 t.u.. The worker moves along the line, so that when product M1 is finished, the worker is at the end of the station, at position 5 l.u.. Although the station has an equivalent length of 5 l.u., the launching interval remains at 3.25 t.u. That means that the second product (M2) enters the station when M1 is at position 3.25 l.u. (the conveyor speed is set to 1 l.u/t.u.). A longer station length results in the possibility of more than one product being in the station at the same time. When the worker finishes M1 at position 5 l.u., M2 is already at position 1.75 l.u.. After finishing the processing of a product, the worker returns along the line with no additional time to start processing the next product. This is illustrated as an arrow between the end position of each product to the start of the next one in the figure. This way, a station length longer than the cycle time can implement flexibility to allow producing multiple product models.

Increasing the length of stations induces, however, costs, such as the equipment, conveyor belt, or the cost of space. There are, therefore, economical and technical limits on how long a station can stretch. It is important to notice that even for stations longer than the cycle time, production infeasibilities may occur. Figure 2.8 presents a chart with the position of the worker in station 1 for a different order of products: M1, M4, M2, and M3. With this new sequence, the length of 5 l.u. is not enough for the production. A black-colored block containing part of the production time of M4 illustrates a station border violation. This violation should be corrected by applying remedial actions, such as stopping the line or assigning a temporary extra worker to

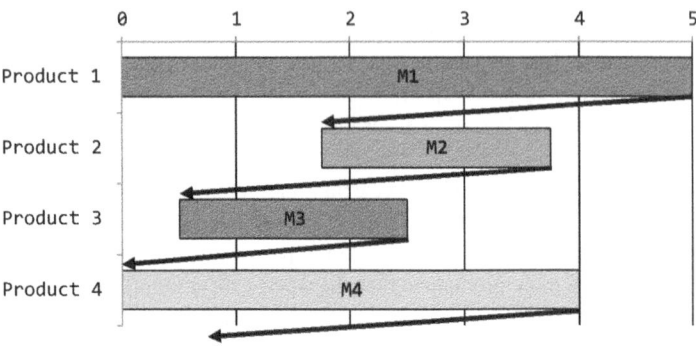

Figure 2.7 Chart with the position of a worker in a station of a paced line

the station. A survey by Boysen et al. [2009c] contains a list of the remedial actions explored in the literature.

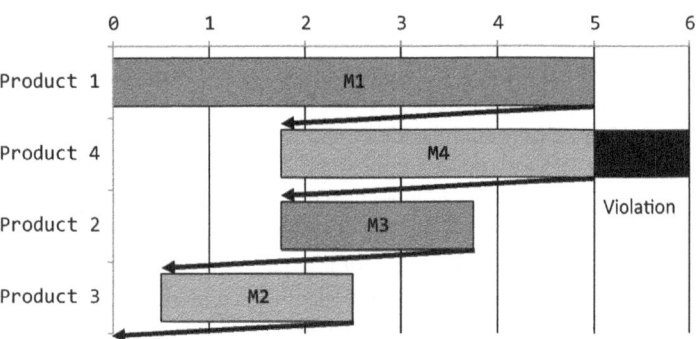

Figure 2.8 Chart with the position of a worker in a station of a paced line with remedial actions

The remedial actions are distinguished between the ones which correct the violation within the station, and others compensate the violation outside the station. Starting within a station, the extra temporary workers who may aid the production are called *utility workers*, which perform *utility work* [Yano and Rachamadugu, 1991]. A utility worker is more costly than a regular worker because he or she must be able to perform the tasks of all stations he or she may support. A second internal compensation is increasing the production speed. A worker can temporarily accelerate the processing of some tasks to avoid a border violation at the cost of a possible lower product quality Boysen et al. [2009c]. The second group of remedial actions compensates for violations within the assembly line. Here, a possible action is to stop the conveyor system until the delayed product is finished in the station Silverman and Carter [1986]. The non-bottleneck stations, however, have increased idle times in this case. Flexible or open station borders are also another possibility. This approach allows workers to extend the stations' borders to assure the production is completed [Thomopoulos, 1967]. Disadvantages are the interactions of workers, who may block themselves and generate idle time. The third class of actions is performed off the assembly line. Assembly lines can have a repair shop at the end of the line, in which non-finished tasks can be performed if this is possible [Kao, 1976]. Alternatively, the repair shop may be installed along the assembly line, so that the product can be reinserted after the tasks are completed. Finally, another remedial action is to either dispose the non-completed product or to account for incompletion costs (loss of value due to lacking features in the product). These remedial actions are normally associated with paced lines. The variable cycle time of unpaced lines has the flexibility to accommodate product variations.

The choice of the cycle time in paced lines presents a trade-off. On the one hand, shorter cycle times exhibit higher productivity. On the other hand, a short cycle time requires more remedial actions to compensate for the deviations, since the shorter the cycle time, the higher is the number of product models that require longer than the cycle time in a station. The best solution depends on the available remedial actions and their costs.

The effect of multiple products also depends on their sequence. As it can be observed by comparing Figs. 2.7 and 2.8, the sequence is essential to determine whether a violation occurs for a given cycle time. Sequences of complex product models in a row (as M1 and M4 in Figure 2.8) are the cause for remedial action, while alternating light (low total processing time) and heavy (high total processing time) products may compensate the processing time variations among themselves. The sequencing in assembly lines is explored in the next section.

2.5   Sequencing in the Automotive Industry

As mentioned previously, the efficiency of automotive production depends on the sequence of the different products. Boysen et al. [2009c] present three main factors as objectives of selecting a suitable sequence: set-up times, due dates, and balancing of assembly lines.

The painting process may present restrictions concerning the sequence of product colors. Some systems require a cleaning time when a new color is selected. Therefore, a suitable sequence may reduce the number of set-ups [Spieckermann et al., 2004; Gagné et al., 2006]. Set-ups between product products may also be present in the body-in-white phase or the final assembly, but most of the literature focus on the paint shop or combinations of the paint shop and final assembly [Boysen et al., 2009c].

Due dates are especially important if production is based on the orders in an assembly-to-order environment [Boysen et al., 2009c]. The production order is, however, not totally fixed by the order in which the products are ordered. The demand may be aggregated in daily or weekly plans which are then sequenced based on other objectives.

The effect of the sequence in the efficiency of the balancing is introduced in Section 2.4. The example of Figs. 2.7 and 2.8 illustrates that the cost of remedial actions (or the productivity itself for unpaced lines) depends on the product order. In the literature, two approaches are used to optimize the sequence aiming at a smooth distribution of heavily loaded products [Boysen et al., 2009c].

The first approach is called *mixed-model sequencing* and corresponds to the sequencing and detailed scheduling of products in the assembly line [Boysen et al., 2009c]. One example of a detailed scheduling can be found in Figure 2.7, in which the start and end production times of each product are defined. The detailed schedule is used to calculate whether the products are assembled inside station bounds and, if needed, the required remedial action for the production in paced lines. For unpaced lines, the mixed-model sequencing is used to determine all production, starvation, and blockage times. McCormick et al. [1989] show that the mixed-model sequencing for unpaced lines is NP-Complete for a fixed number of stations. For paced lines, Yano and Rachamadugu [1991] present an analytic solution for the case of one station and two product models. For more product models, Yano and Rachamadugu [1991] develop only a heuristic.

A second approach simplifies the sequencing and does not require a detailed scheduling of pieces. The *car-sequencing problem* aims at finding a sequence of products that respect some sequencing rules [Drexl and Kimms, 2001]. These rules are mostly modeled as X out of Y products. That is, in a sequence of Y products, at most X products containing a given characteristic may be assigned [Fliedner and Boysen, 2008]. An example of such a characteristic or option is the presence of sunroofs. Their installation may require a significant processing time in a station, while a car without a sunroof does not need this task at all. Therefore, a sequence rule such as at most 1 product may have a sunroof in any sequence of 3 vehicles is sensible. The sum of all rules in all stations defines the car-sequencing problem.

The objective is to find a sequence that obeys all rules or that minimizes the number of disregarded rules [Fliedner and Boysen, 2008].

Both the mixed-model sequencing and the car-sequencing formulations are applied to the same class of problems. There are, however, significant differences between the two approaches. The advantages of car-sequencing are based on its simplicity. This approach is easier to understand and to implement so that sequencing rules are broadly used in the industry [Golle et al., 2014a]. The car-sequencing, however, has also a number of downsides. One of them is the complexity of generating the sequencing rules. Although a 1:3 rule is easy to understand, enumerating the exact rules for all combinations may be a very complex problem. Furthermore, these rules consider that there are only two possible product model options in each station [Golle et al., 2014a]. Cases in which three or more processing time alternatives occur in one station cannot be modeled with such rules. An example with 3 product models (M1, M2, and M3), in which product model M1 has a processing time below the cycle time, product model M2 has slightly a larger processing time than the cycle time, and product model M3 a longer processing time is already enough to show the difficulty of applying such rules. For the combinations of product model M1 and either only M2 or M3, simple rules such as 2:3 (for product model M2) or 1:3 (for product model M3) would suffice. A production sequence M2-M2-M3, however, respects both rules (there are at most 2 instances of product model M2 and 1 of product model M3 within a sequence of 3 products) but would definitively require some remedial action. The third disadvantage of car-sequencing is that its objective is only a proxy for the minimization of the use of remedial actions [Golle et al., 2014a]. If the optimal answer of an instance must disrespect one or more rules (otherwise it would be infeasible), it does not necessarily mean, that this sequence requires the minimal amount of remedial actions. Golle et al. [2014a] show that optimal answers of the car-sequencing problem minimizing rule violations require at least 15% more utility work comparing to the mixed-model sequencing solution, which directly minimizes utility work. Furthermore, Golle et al. [2014a] discuss that computational tests show that car-sequencing is only significantly easier and faster to solve for simple rules and under some further assumptions.

Another cost that is affected by the sequencing is the part-feeding logistic. Different products require specific parts which must be available at the station for the assembly. Therefore, a possible sequencing objective is to smoothen the part usage to facilitate the logistic process [Miltenburg, 1989]. If product models are evenly spread with respect to the parts they need, the storage of components in each station would be used at a constant rate. Under known and controlled usage rates, the logistic process may be simplified and yield lower inventory levels at the stations. In the literature, this problem is named *level scheduling* [Boysen et al., 2009c].

Often a sequence must incorporate several if not all of the elements: set-up, due dates, processing time variations, and part-feeding. The linear design of the production restricts the reordering of products so that the sequence at the beginning of the process in the body-in-white is very similar to the output sequence of the final assembly. According to Boysen et al. [2009c], some companies set up an interlinked conveyor system, so that the sequence must remain the same in the whole production process. Buffers (described in Sec. 2.6) can provide some flexibility, although their application is limited due to the product sizes [Lopes et al., 2020b]. Often the multiple sequencing objectives are conflicting. Drexl and Kimms 2001, for instance, solve the sequencing considering simultaneously station load and part levels.

The sequencing of automotive production in a real application is a hard problem to solve. Not only several objective functions may be modeled at the same time, but also the instance sizes can be very large. The master scheduling of the weekly production, for instance, has to sequence more than 10,000 products considering a continuous production of a vehicle per minute. Therefore, heuristics and rules are still broadly in use [Golle et al., 2014b].

These difficulties also translate to the balancing problem. Both, balancing and sequencing problems, are closely related, since each task assignment may determine whether a product model requires a high or low processing time in that station. Efficient lines not only have a good quality balancing and sequencing, but also an assignment that creates synergies between them. These effects are hard to combine since balancing and sequencing have a different decision time frame. The balancing problem is a design problem. Once implemented, the assembly line has a lifetime of months to years. On the other hand, sequencing is mostly a short-term decision. The production quantities are determined usually by the product orders, so that each week or day may have a different production sequence. Therefore, a hierarchical approach may be considered, in which the balancing is solved before a subsequent sequencing optimization Scholl [1999]. Lopes et al. [2020b] test the inverted hierarchical approach by optimizing the balancing for known sequencing configurations with improvements in the order of 5% for the given dataset. Therefore, it is meaningful to combine the effects of both problems. Exact solution methods for the simultaneous balancing and sequencing problem appear in the works of Karabati and Sayın [2003]; Öztürk et al. [2015]; Lopes, Michels, Sikora, Molina and Magatão [2018]; Lopes et al. [2019]. However, these approaches are usually limited to small sequences, due to the required computational effort to solve the problem. Furthermore, a unique sequence is generally considered, while in practical applications the balancing has to cope with a variety of product sequences.

2.6   Buffers

The interrelationship between sequencing and balancing and the severe limitations on sequence control in the automotive industry are described in the last section (Sec. 2.5). Due to conveyor belt systems and the large size of products, reordering vehicles is not always possible. Such a reordering is, however, not impracticable. There are some stages of the production in which buffers are often installed [Boysen et al., 2011]. Such buffers do not only compensate production level discrepancies and failures but also allow for a local reordering or resequencing of the products.

Buffers may differ greatly between different industries: the products of the electronic industry, for instance, are small and can be stored in boxes while in the automotive industry, a product piece is a car or a truck. Not only the opportunity cost of storing an electronic chip and a truck differ but also the handling system to support the buffer operation depends on the application. While electronic components can be stored manually, a worker may not be able to move a truck without aid. In the automotive industry, most of the buffer usage exists between the different stages of the process (body-in-white, paint-shop, and final assembly) [Boysen et al., 2009c]. The operations of buffers at the paint shop are described by Spieckermann et al. [2004]. These buffers are commonly used to change the order of the products based on their color. After the paint-shop, another buffer may exist to decouple the sequence before the final assembly. These buffers, however, may be only large enough for units or dozens of units. The resequencing may also be limited according to the buffer organization.

Three types of buffers are found in the balancing literature: pull-off tables, mix banks, and automated storage and retrieval system (AS/RS) [Boysen et al., 2011]. Pull-off tables are the most simple buffers and are usually small in size [Boysen et al., 2011]. As the name describes, this buffer system consists of space at the side of the conveyor belt and a handling system that can move products from the conveyor to the buffer or vice-versa. Figure 2.9 illustrates two examples of the options such a buffer has to offer. Products are moved along a conveyor from left to right. In the top part of the figure, one of the two slots of the buffer is occupied with product 2. The available options are returning product 2 to the line, assigning product 3 to the second spot of the buffer, or send product 3 forward to production after product 1. The bottom part of the picture brings an example, in which the buffer is fully occupied with products 2 and 3. The options for selecting the next product are either to release products 2 or 3 from the buffer or to keep them and pass along product 4. This buffer structure is modeled in a car resequencing problem in Boysen et al. [2011].

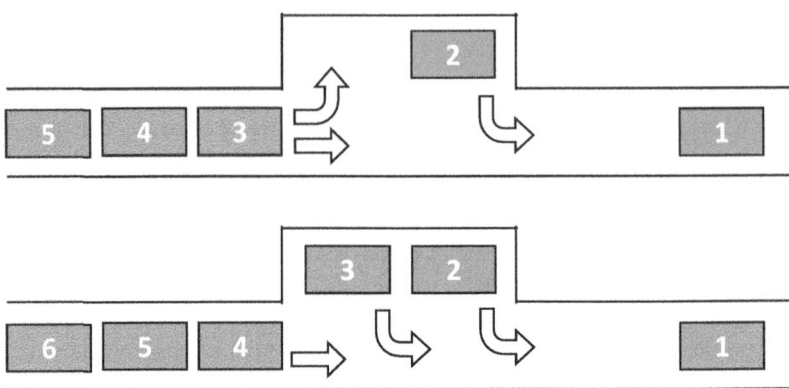

Figure 2.9 Pull-off table buffer

A second buffer configuration is the mix bank buffer. In this system, the entry conveyor belt is connected to several parallel lanes [Meissner, 2010; Taube and Minner, 2018], as illustrated by Figure 2.10. At the entry of a product, the system must assign it to one of the lanes. Each lane has a fixed number of positions, in which products cannot overtake the others. At given intervals, one of the products at the head of the lanes is chosen to be passed along to production. This buffer structure allows a degree of reorganization of the products, which is however limited by the non-overtaking restriction.

The third buffer structure is called *automated storage and retrieval system* (AS/RS) [Meissner, 2010]. This system is the most versatile since any product can be stored in and retrieved from any position at any time. A graphical illustration is found in Figure 2.11. The AS/RS works with independent slots, which can be filled or retrieved without any restrictions. An AS/RS with N positions can be seen as a mix bank buffer with N single-capacity lanes. The efficiency of mixed banks and AS/RS are compared in the sequencing context by Meissner [2010].

Buffers do not only differ in structure, but their use can also aim at different objectives. As automotive companies rely heavily on long supply chains with rigid logistic planning, a common reordering objective is to keep the planned sequence. This way, customized parts can be delivered within the planned horizon reducing inventory costs. Therefore, buffers are often used to return products to the initial sequence if any errors or failures occur. Examples of this resequencing objective are found in Meissner [2010] and Boysen et al. [2011].

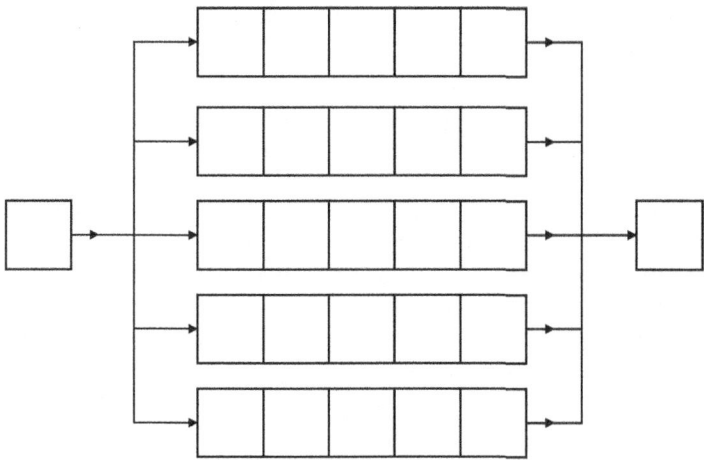

Figure 2.10 Mix bank buffer

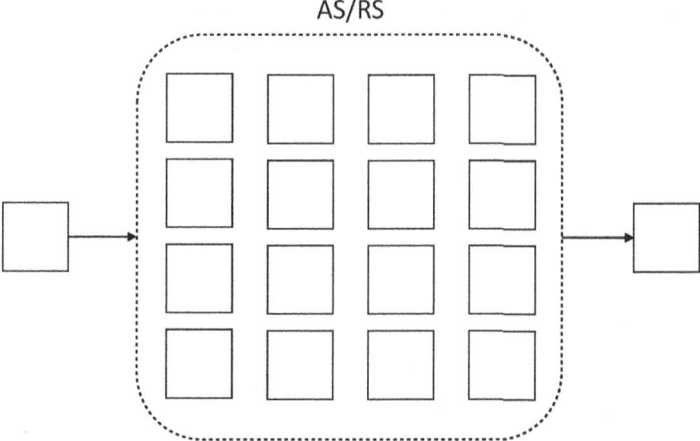

Figure 2.11 Automated storage and retrieval system (AS/RS) buffer

A buffer system can also be used with other objectives, such as reducing the cost of remedial actions, improving the part-level scheduling, or decreasing the set-up times in the paint-shop. For instance, Lopes et al. [2020a] model the balancing, sequencing, and buffer allocation as an integrated problem to maximize the productivity of unpaced lines. Taube and Minner [2018] model the resequencing problem for the final assembly. They divide the assembly process into two parts: a first in which sequences can be altered, and a second, in which products must respect the master sequencing. The approach optimizes the sequence for multiple objectives (utility work, set-ups, and part leveling) with the restriction that the reordering must be reversible to the original sequence.

One interesting development on the future design of assembly lines is the use of automated guided vehicles (AGV) for the transport of the pieces. According to Kampker et al. [2017], AGVs make routing between stations possible, so that overtaking or non-linear production sequences become possible. This would increase greatly the resequencing capabilities of the automotive industry. The optimization of a flexible layout for the automotive industry is explored by Hottenrott and Grunow [2019]. The authors solve the balancing problem along with an AGV flow model to reduce the traveled distances. These examples show how complex the balancing of assembly lines is and how new technologies provide more flexibility for the design of production systems.

Literature Review on Assembly Line Balancing Under Uncertainty

3

The literature on the assembly line balancing problem under various forms of uncertainty is explored in this chapter. Firstly, the surveys on assembly line balancing and related problems are introduced and their classification structure is explained. The literature is then described based on its source of uncertainty. Finally, the contributions of this work are described and aligned with the shown gaps in the literature.

3.1 Scope and Structure

The scope of the literature analysis in this chapter is to provide a classification and overview of papers published on assembly line balancing, sequencing, and buffer allocation that have any uncertainty in its formulation. The review build itself up from the excellent reviews in balancing [Boysen et al., 2007, 2008; Battaïa and Dolgui, 2013], sequencing [Boysen et al., 2009c], resequencing [Boysen et al., 2012], disassembly [Özceylan et al., 2019], and buffer allocation [Weiss et al., 2019]. The existing surveys in the literature are discussed in Section 3.2.

For the review of the literature, a classification scheme is introduced in Section 3.3. This classification is divided in the following section into uncertainties in single-model production systems (Section 3.4), uncertainties in multiple-model production systems (Section 3.5) and uncertainties in disassembly systems (Section 3.6).

The section on single-product systems contains a review of the assembly line balancing problem, in which stochastic and fuzzy processing times are explored. Furthermore, a section on buffer allocation explores the literature that combines processing time allocation with buffer placing.

The uncertainties on multiple-model production systems are divided into approaches modeling uncertainty mainly in stochastic processing time and other

© The Author(s), under exclusive license to Springer Fachmedien Wiesbaden GmbH, part of Springer Nature 2022
C. G. Stall Sikora, *Assembly-Line Balancing under Demand Uncertainty*, Gabler Theses, https://doi.org/10.1007/978-3-658-36282-9_3

less popular uncertainties, such as uncertainty in the production sequencing and the demand.

Section 3.6 is devoted to the uncertainties of disassembly processes. The disassembly process is not simply an assembly in the reverse order and may be affected by several sources of uncertainty. As the products are in their end-of-life, the quality and integrity of the components, for instance, is strongly uncertain.

3.2   Existing Surveys in the Literature

There are multiple surveys in the literature covering several aspects of the design and operation of production systems.

Boysen et al. [2009b] present an overview on production planning of mixed-model assembly lines in which several levels of planning are discussed. Boysen et al. [2009b] divide the decisions related to assembly lines into assembly line balancing, master sequence planning, rebalancing, sequencing, and resequencing. These problems are mostly solved independently, but approaches that integrate multiple levels or solve them in a hierarchical approach [Meyr, 2004] may present better efficiency [Boysen et al. 2009b]. This integration is however hard to enforce since the different levels of decisions require information that is available at different steps of the planning process.

This section points out the existing surveys to the related problems of assembly line design and operation. The main focus is on the sources of uncertainty and the problems' stochastic variants. The surveys are divided into the five levels introduced by [Boysen et al., 2009b] (balancing, master scheduling, rebalancing, sequencing, and resequencing) plus the topic of buffer allocation, which is also important in production systems under uncertainty.

3.2.1   Balancing

Since first introduced in a scientific paper in 1955 by Salveson [1955], the assembly line balancing problem has interested several researchers and stimulates a highly active field until the present days. The amount of surveys devoted to the topic illustrates well the importance of the field and the vast number of publications. A non-extensive list of review papers is composed by Baybars [1986], Ghosh and Gagnon [1989], Gagnon and Ghosh [1991], Erel and Sarin [1998], Becker and Scholl [2006], Scholl and Becker [2006], Boysen et al. [2007], Boysen et al. [2008], Battaïa and Dolgui [2013], Bentaha et al. [2015], Özceylan et al. [2019], and Eghtesadifard et al. [2020].

The broad literature justifies survey papers to specify a subset of the problem treated, such as the simple version of the problem (SALBP) [Baybars, 1986; Scholl and Becker, 2006], heuristic procedures [Erel and Sarin, 1998], stochastic effects [Bentaha et al., 2015], and many others. An excellent classification of problems is given by Boysen et al. [2007], while Eghtesadifard et al. [2020] provide a systematic review of assembly line problem trends.

Due to the extension and number of surveys on the balancing problem, only the stochastic or uncertain characteristics discussed in the reviews are examined in this section. A summary of the listed surveys on assembly line balancing problem is given in Table 3.1 along with their scope. Besides Bentaha et al. [2015], no further survey focuses on the uncertainty of assembly-line problems.

Table 3.1 Summary of existing surveys about assembly line balancing and their scope

Author(s)	Scope
Baybars [1986]	Solution methods for the simple assembly line balancing problem
Ghosh and Gagnon [1989]	Review and analysis of assembly line balancing problem, along with assembly line design considerations
Gagnon and Ghosh [1991]	Relation between research on assembly line balancing problem and the industrial application of methods
Erel and Sarin [1998]	Heuristic methods for the assembly line balancing problem
Scholl and Becker [2006]	Solution methods for the simple assembly line balancing problem
Becker and Scholl [2006]	Solution methods for the generalized assembly line balancing problem
Boysen et al. [2007]	Classification of assembly line balancing problems
Boysen et al. [2008]	Assembly line balancing models and applications
Battaïa and Dolgui [2013]	Comprehensive review of assembly line balancing from 2007-2012
Bentaha et al. [(2015]	Assembly line balancing with stochastic processing times and disassembly line balancing
Özceylan et al. [2019]	Disassembly line balancing
Eghtesadifard et al. [2020]	Systematic review on assembly line balancing

Although the surveys differ greatly in their scope, date, and depth, they all have in common the strong association of uncertainty with the processing times. Among the surveys, the distinction of deterministic and stochastic processing times is a common criterion to sort out the literature, as in Battaïa and Dolgui [2013]. The classification of Boysen et al. [2007] contains the class "t^{sto}" for stochastic processing times, which is one of the two only mentions of uncertainty in the classification scheme. The second one is related to the cycle time restriction. The class "$prob$" relates to models in which the cycle time must be respected with a given probability. This restriction is useful to model stochastic processing times (as a chance-constraint) or for mixed-model assembly lines, in which a given proportion of models should be produced within the cycle time. This section reviews the surveys' insights on stochastic processing times and other less mentioned uncertainty factors, such as disassembly processes and the production system costs.

Stochastic processing times are usually related to manual assembly lines. The instability in the work rate is justified by the different levels of motivation, skill, and fatigue of workers or failures in machines or the logistic system [Becker and Scholl, 2006]. According to Becker and Scholl [2006], the variation of the processing times is strongly related to the task complexity. The uncertain processing times are modeled as random variables (mostly under the assumption of a normal distribution), as fuzzy numbers, within given intervals, or by scenario definitions [Battaïa and Dolgui 2013]. Along with stochastic optimization, robust optimization is also an explored modeling alternative in the assembly line balancing literature [Bentaha et al., 2015].

One important criterion for the classification of the literature dealing with stochastic processing times is how the variability is addressed. Optimization approaches can either minimize the effect of uncertain processing times by distributing the variation equally between stations or by assigning tasks so that the probability of a very loaded station is minimized. Alternatively, the economical cost of the response to the variation can be accounted for and integrated into the optimization. Figure 3.1 contains a classification of the usual approaches to cope with processing-time variations. The different approaches depend on the available data and how detailed the modeling of the assembly system is. Some responses, such as employing utility work, may also require the detailed scheduling of workers. For this response, not only the balance of the line but also the realization of the processing times must be accounted for. The collection of reactions to processing time variations is listed in a survey of the sequencing of mixed-model assembly lines by Boysen et al. [2009c]. Processing time variations can be caused by stochastic processing times or by the presence of multiple products. Although both causes

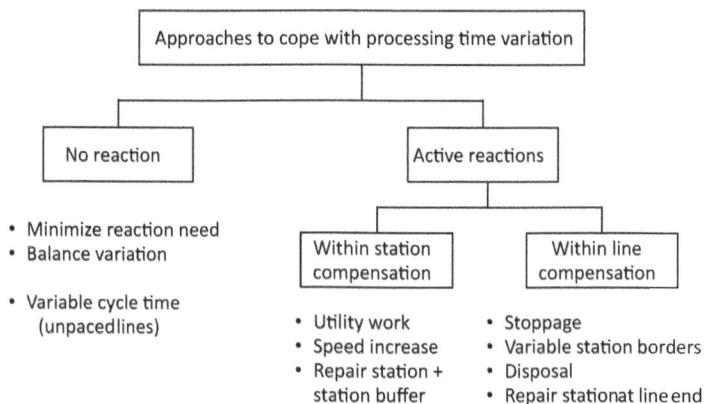

Figure 3.1 Classification of the reaction approaches to cope with processing time variation

are different, the modeling approaches to deal with the resulting variability are the same.

The approaches to deal with the processing variation listed in figure 3.1 can be divided into two major groups, the ones that do not consider a reaction and the ones that compute the needed reaction. Within the models that do not model a reaction, the first possibility is to minimize the effect of the variation using a proxy for the reaction. One example is given by Kao [1976], who considers the probability of a task assignment surpassing the cycle time and aims at its minimization. The incompletion costs are not dealt with, instead, an assignment is looked for which exhibits a high completion probability. A second possibility is to even out the variation between the stations. In the context of stochastic processing times, Raouf and Tsui [1982] aim at the minimization of the stochastic variation in the assignments. The same approach can be seen in the context of mixed-model assembly lines in Thomopoulos [1970], in which the variation between the products is minimized in a horizontal balancing approach. A third possibility occurs in unpaced assembly lines. As the pieces move forward only when the operations are finished, the cycle time is variable. The handling system itself reacts accordingly to the processing-time variation. All the other reaction possibilities refer to paced assembly lines, although it would be also possible to implement some of the other compensations in unpaced lines as well.

The active reaction to processing-time variation is divided into whether the effect is dealt with within a station or among multiple stations (or the whole line). The reactions within a station are the use of utility work, controlling the worker speed, and repair shops. Utility work is auxiliary work that is performed by all-around workers that aid the processing when needed [Tsai, 1995]. Utility workers are more costly than regular workers so that their additional use is to be minimized. There is a trade-off between the cost of the utility work used and the sizing of the system. A second option is to use the worker's speed as a reaction. It is possible to operate temporarily at a higher pace so that a product with a larger processing time can still be finished within the cycle time. Higher speeds, however, result in lower product quality, so the trade-off between the costs of the system and the losses in quality must be quantified [Becker and Scholl, 2006]. The third approach within stations is the use of repair shops along the line. If a piece is not completed, it is sent to the repair shop and a piece previously placed in a buffer takes its place [Buzacott, 1999]. When it is finished, the order is placed in the buffer. In these three approaches, the variation is dealt with before the piece arrives in the next station so that the reactions do not affect any other station of the line.

There are also measures that influence more than one workstation. The first alternative is to stop the whole line if any station needs more than the cycle time for the processing of its product [Tsai, 1995]. If workers are allowed to exceed the station boundaries, flexible or variable station bonders can be used [Dar-El and Nadivi, 1981]. This reaction gives flexibility to cope with large processing times but may require extra coordination between workers. As stations' borders are not fixed in this case, one worker may block the way of another worker. Disposal of the non-finished workpieces is also an alternative. In this case, lost material and work costs must be accounted for. Finally, there is also the case of repair stations along the line or at the end of the line [Kottas and Lau, 1973]. These stations are responsible for the completion of the tasks that could not be completed within the cycle time. The difference between the "repair station"-reaction classified as a "within station compensation" lays in the effect on other stations. If the repair is made only at the end of the line, the tasks which depend on the delayed task cannot be performed in later stations. Similarly, the use of repair shops along the line without the use of buffers may cause empty cycles in later stations. In summary, these four last approaches do not react to the processing time variation within only one station. Delays in one station cause consequences in the later stations of the line.

There are, however, several arguments against the use of uncertain processing times in assembly line balancing models. According to Falkenauer [2005], assembly lines apply very specialized work for small subsets of tasks, so that the production time is mostly deterministic. Furthermore, Boysen et al. [2008] state that in the automotive industry, the processing times are agreed upon with the workers' union and are stable for skilled workers. There is also literature that considers the processing time of workers in training in form of learning effects Boysen et al., [2008]. In these cases, the processing time is considered dynamic Boysen et al., [2007].

A different source of uncertainty in the balancing of assembly lines occurs when the inverse process is performed: the product disassembly. Disassembling products is prone to other uncertainties due to the state and quality of already used products. The survey of Özceylan et al [2019] on the balancing of disassembly lines points out that not only the processing times can be stochastic, but also the quality of the pieces and the number of subparts. Furthermore, the reliability of a disassembly process is lower than that of an assembly process. Parts can get damaged or broken during the process. Bentaha et al. [2015] also survey the balancing of disassembly lines. The focus of this survey is on the stochastic processing times in assembly and disassembly lines.

The third source of uncertainty is only mentioned in the surveys, but is not explored in any classification or has any dedicated literature niche. Becker and Scholl [2006] state that costs and profits from assembly lines are difficult to estimate. The normally used objectives such as number of stations or cycle time are just a proxy of the given costs or profits. There is no publication of the assembly line balancing problem that focuses explicitly on the uncertainty on equipment or design-related costs.

3.2.2 Master Scheduling

The master scheduling links the long-term configuration decisions of a production system to the short-term decision of assigning the individual orders to producing periods. According to Boysen et al. [2009b], the master scheduling of the automotive industry considers time horizons with lengths of about a month and is used to decide the production of individual shifts. According to the survey on production planning of mixed-model assembly lines by Boysen et al. [2009b], the topic is hardly explored in the literature.

The master scheduling objective is to plan the production of the individual customer orders. Each order has a due date, which may cause a penalty if the delivery is delayed. Analogously, producing too early induces holding costs. The order assignments must also consider the production restrictions, such as time and resource availability. Boysen et al. [2009b] state that in practice, the master scheduling is mainly dominated by the marketing and sales department, which may assign the production orders in the order the products are sold.

Boysen et al. [2009b] present a basic model for the master scheduling along with extensions such as capacity and demand adjustments. Demand adjustments consist of the possibility of rejecting orders and renegotiating due dates. The production capacity can be altered by an assembly line reconfiguration or rebalancing, by adjusting the number of shifts and their lengths, or by subcontracting.

The master scheduling problem is modeled by Boysen et al. [2009b] as a deterministic problem. The authors state, however, that a rolling horizon scheme is more adequate for the implementation of the solutions. This way, only one shift or day is implemented at a time and the others are reoptimized considering the new sales information that may become available.

3.2.3 Rebalancing

There are no survey papers on the rebalancing of assembly lines yet. For more information about the literature, Gamberini et al. [2006] or the assembly line balancing surveys of Battaïa and Dolgui [2013] and Eghtesadifard et al. [2020] are recommended.

The rebalancing of assembly lines deals with a trigger such as a change in cycle time, demand mix, or product variation and may have restrictions due to the commitment of a previous assembly line implementation. Among the uncertainties in solving such a problem, the correct estimation of moving or adjusting cost can be of interest. Furthermore, the variability of the future demand may also be a source of uncertainty.

3.2.4 Sequence Planning or Sequencing

The sequencing problem occurs in assembly lines with multiple models, in which the order of the products is important for the cost or productivity calculation. According to the decision levels of Boysen et al. [2009b], the sequencing of products is an operational decision that is performed every day or at every production shift.

Boysen et al. [2009c] present a survey on sequencing approaches applied to mixed-model assembly lines. The authors identify three major approaches to sequence the production in assembly lines: mixed-model sequencing, car sequencing, and level scheduling. The mixed-model sequencing models detailed processing times and worker positions for a sequence of models. More and less loaded products are intercalated to compensate for each other. The car sequencing is a less detailed variant of the workload modeling approach, in which the sequences are restricted with simple rules. These rules have an H:N form, which means that at most H models with a certain characteristic may be present in a sequence of N products. A third approach is the level scheduling, which aims at balancing part consumption. A stable part usage goes along with the Just-in-Time supply chains that are typically encountered in the automotive industry. Balanced consumption levels result in lower inventory levels and are used as a proxy for the material supply costs.

The approaches surveyed by Boysen et al. [2009c] concentrate on paced lines without buffers with deterministic and known demand. In the whole survey, only one reference [Chutima and Yiangkamolsing 2003] deals with non-deterministic data, in the form of fuzzy processing times.

3.2.5 Resequencing

The resequencing problem consists of changing the sequence of products within the production process. The resequencing in mixed-model assembly lines is surveyed by Boysen et al. [2012], who present a classification scheme based on five factors.

The first factor is the trigger of the resequencing. Boysen et al. [2012] divide the literature between reactive and proactive approaches. Reactive approaches aim to correct or improve the sequence after a disturbance occurred. These disturbances can be machine failures, delays in the part material deliveries, production or quality errors, etc. Proactive resequencing, on the other hand, uses the reordering flexibility according to the multiple sequencing objectives along the processing process. The proactive approaches are usually applied in the boundaries of the body-in-white and paint-shop or paint-shop and final assembly so that a better sequence for each individual phase can be achieved.

The buffer configuration is the second criterion in the classification. Buffers can be implemented as pull-off tables, mix banks, or AS/RS systems as described in Section 2.6. Furthermore, Boysen et al. [2012] also consider virtual resequencing, which does not use buffers. The virtual variant reassigns physical products to customer orders. If a specific component for an order is not delivered in time, but a second order is similar to the product, the physical product can be assigned to the second order. The first order is then delayed and produced later.

The third classification criterion is the problem's decision variables. Firstly, the problem can be defined as an installation or operation problem. Installation problems focus on the placing, sizing, and selecting of the buffer system. Operational problems optimize the sequence itself and are further divided into static and online problems. Static problems present all the information beforehand, while online problems are dynamic and require decisions without full knowledge of the future sequence.

The different objective functions are described as the fourth classification criterion. One common objective is the restoration of the planned sequence after perturbations. Another approach is to use an objective function of a sequencing problem. The listed related problems are paint batching, mixed-model sequencing, car sequencing, and level scheduling [Boysen et al. 2012]. Finally, the fifth criterion is the solution method. Boysen et al. [2012] divide the literature into exact, heuristic, and simulation methods.

The resequencing problem is a response to uncertain events that disrupt the planned production sequence. This problem can be either modeled as a deterministic response given all information of the sequence and disruption or as a stochastic or online problem [Boysen et al., 2012]. When the information is incomplete, uncertain, or there is not enough time to consider the whole system, an online approach is more appropriate.

3.2.6   Buffer Allocation

The literature on buffer allocation is only partly related to the literature on assembly line balancing or sequencing. Much of the literature on assembly lines considers paced lines and no buffers, while most of the literature on buffer allocation considers the production system to be given. Considering the differences in methods and applications, the surveys on buffer allocation are described in more detail in this section among their characteristics and the classification of the buffer allocation literature.

The sizing and placing of buffers in the literature is named the *Buffer Allocation Problem* (BAP). Buffers are usually used to compensate stochastic factors by decoupling the material flow, although they can also be used for some deterministic compensations. The literature on buffer allocation is usually clustered in whether the production system is balanced and in whether the machines are reliable [Demir et al., 2014].

A balanced production system has all its machines operating at the same rate. That is, their processing times are identical, if they are deterministic, or they have the same mean if they are taken as stochastic [Demir et al., 2014]. An unbalanced system presents workstations with larger expected processing times than the average. Hillier and Boling [1979] show that the imbalance can increase production levels, if the larger processing times are concentrated at the beginning and the end of the line, leaving the center of the line with smaller expected processing times. Because of the shape of the processing-time curve based on stations, this effect is named bowl phenomenon [Hillier and Boling, 1979]. One of the first surveys on the topic is published by Smunt and Perkins [1985] in which the effects of imbalance on unpaced stochastic lines are discussed. Smunt and Perkins [1985] state that the bowl phenomenon appears only situationally, mainly in short lines with high processing-time variance. A more recent survey on the unbalanced effects is given by Hudson et al. [2015]. This latter survey points out that the imbalance (or the bowl phenomenon) can occur with the mean values of the processing times or with the variations of the processing times as well.

Another important classification element for buffer allocation problems is the reliability of machines. An unreliable machine may break down and may need repair before resuming the production activities. Buffers before a machine can accumulate workpieces of previous machines so that they are not blocked during the repair. Similarly, an empty buffer after the machine assures that the next machines can still temporarily work on the workpieces stored in the buffer.

In 2000, Gershwin and Schor [2000] present a survey on solution methods for the buffer allocation problem. The authors classify the literature based on the different objective functions, whether the models are continuous or discrete, and the used solution method.

The survey from Demir et al. [2014] extends the survey from Gershwin and Schor [2000], covering articles from 1998 up to 2012. Demir et al. [2014] propose a more formal classification scheme, in which papers are classified based on the machine reliability, the topology of the production system, the solution methodology, and the objective function. In the considered period, Demir et al. [2014] gather 41 references optimizing reliable systems, while 54 focus on unreliable machines. Among the unreliable systems, the work on quality inspection from Han and Park [2002] is also considered. Demir et al. [2014] classify the system topology into serial lines, parallel lines, general networks, assembly lines, flexible manufacturing systems (FMS), and cellular manufacturing systems (CMS). Demir et al. [2014] point out that solution methods are usually formed by a generative and an evaluative method. Evaluative methods are responsible for calculating or estimating the value of a given buffer configuration. The evaluative methods are divided into analytic and simulation-based procedures. According to Demir et al. [2014], analytic methods can be both exact or approximate. The exact methods are limited to very small lines, while the approximate methods aim at decomposition and/or aggregation of small systems into larger ones or vice-versa. Even though larger lines can be solved this way, analytical methods are restricted to strong assumptions Demir et al. [2014]. These assumptions generally are deterministic or exponentially distributed processing times and geometric or exponential failure rates. The generative methods are used to propose buffer assignments for a test with the evaluative methods. Demir et al. [2014] point out that enumerative methods are limited to very small problems, while metaheuristics dominate the literature for larger problems. Finally, based on the objective function, articles are divided into production maximization, buffer size minimization, system cost minimization, profit maximization, work-in-process (WIP) minimization, and other objectives.

The most extensive and up-to-date survey on the buffer allocation problem is given by Weiss et al. [2019]. In this survey, the literature is classified in detail based on the characteristics of the flow line, its objective functions and constraints, and the employed solution method. Weiss et al. [2019] survey only papers dealing with some kind of uncertainty. They point out that the allocation of buffers is an NP-Hard problem, and that analytic solutions can only be found for very small instances under very strict assumptions.

Weiss et al. [2019] extensively detail the differences in flow systems. Their classification takes into account the control mechanism for the transport of pieces, the supply of raw material, the blocking type, the reliability of machines, and whether the system is balanced or unbalanced. The control mechanism is divided into a classic flow system with unitary products and a control system with pallets or skids. The supply of raw materials is classified into saturated or unsaturated. Saturated supply assumes that there is enough raw material at all times at the start of the line. Unsaturated systems operate with an entry buffer and may depend on random arrivals or some order policy (such as (s,q) or (r,S)). The same kind of uncertainty can also be assumed for the output of the line, in form of demand uncertainty. The blocking of machines is distinguished into after station (BAS) and before station (BBS). A blocking after station occurs when the output buffer of a station is full and the station cannot forward the next finished product so that the machine must wait for a vacant buffer. The BBS condition is stricter since a machine can only start its operation when a position in the next buffer is available. Finally, the unreliability of machines is further classified into operation or time-dependent. The failure rate, repair rate, and processing-time distributions are also described in the survey for each one of the references. Most common are the exponential or the Erlang distributions for the processing times of reliable machines while unreliable machines are modeled with deterministic processing times in the majority of papers. The time between failures is mostly considered as geometrically or exponentially distributions.

The classification of objectives in Weiss et al. [2019] is based on five major measurements: throughput, work-in-progress, time-in-system, customer service, and buffer size. Weiss et al. [2019] describe the two most common versions of the problem, the minimization of the buffer size for a given expected productivity level, and the maximization of the expected production for a given buffer size. Furthermore, Weiss et al. [2019] describe the variations of the mentioned restrictions, such as non-blocking probabilities or service level restrictions for the buffer size and throughput problems. The combinations of two or more of the five main objective functions are also described, discussing the trade-offs observed in each reference.

The classification of the solution methods from Weiss et al. [2019] is also more extensive than the one by Demir et al. [2014]. Weiss et al. [2019] divide methods into explicit solutions, integrated optimization methods, and iterative optimization methods. The explicit solutions are either exact analytical methods, analytical methods based on approximations, or methods based on optimality conditions. Similar to Demir et al. [2014], Weiss et al. [2019] state that exact methods are limited to small and restricted systems. The integrated optimization methods rely on either samples of realizations or analytical results for the generation of models. Finally, iterative optimization methods are the ones that can be divided into the generative and the evaluation part, such as described by Demir et al. [2014]. Weiss et al. [2019] classify iterative methods into enumerations, metaheuristics, search algorithms, and dynamic programming. Examples of search algorithms are bottleneck-based search, gradient algorithms, and derivative-free algorithms.

This overview on surveys of buffer allocation problems shows that the buffer allocation is often used to compensate for stochastic effects, mostly stochastic processing times and unreliable machines. In the majority of the reported references, the production system is given and only the buffer is considered a problem variable.

3.3 Classification Scheme

To present the literature in an orderly manner, a classification scheme is proposed. This classification is focused on the modeling of the uncertainty presented in each article. There are also other important factors for contributions on assembly line balancing that are independent of the nature of the uncertainties. Assembly lines may contain set-up times, parallel stations, or present themselves in a U-form. In order not to neglect these characteristics, the presentation of the articles' features also considers the assembly line balancing classification of Boysen et al. [2007]. This classification uses tuples $\{\alpha, \beta, \gamma\}$ to describe the tasks, line, and objective function characteristics, respectively. Within the tuples, acronyms are used to express the presence of a characteristic in the treated problem. The classification scheme is described in Tables 3.2 – 3.4.

The possible values for α are given in Table 3.2 and represent either task or precedence graph characteristics. Table 3.3 contains the classification scheme for the station and line characteristics, expressed using β. The objective functions are expressed by γ and are summarized in Table 3.4.

Table 3.2 The classification scheme of Boysen et al. [2007] for the precedence graph characteristics

Precedence Graph Characteristics	
Product specific precedence graphs	
$\alpha_1 = \text{mix}$	Mixed-model production
$\alpha_1 = \text{mult}$	Multi-model production
$\alpha_1 = \circ$	Single-model production
Structure of the precedence graph	
$\alpha_2 = \text{spec}$	Restriction to a special precedence graph structure
$\alpha_2 = \circ$	Precedence graph can have any acyclic structure
Processing times	
$\alpha_3 = t^{sto}$	Stochastic processing times
$\alpha_3 = t^{dyn}$	Dynamic processing times (e.g. learning effects)
$\alpha_3 = \circ$	Processing times are static and deterministic
Sequence-dependent task time increments	
$\alpha_4 = \Delta t_{dir}$	Caused by direct succession of tasks (e.g. tool change)
$\alpha_4 = \Delta t_{indir}$	Caused by succession of tasks (tasks hinder each other)
$\alpha_4 = \circ$	Sequence-dependent time increments are not considered
Assignment restrictions	
$\alpha_5 = \text{link}$	Linked tasks have to be assigned to the same station
$\alpha_5 = \text{inc}$	Incompatible tasks cannot be combined at a station
$\alpha_5 = \text{cum}$	Cumulative restriction of task-station-assignment
$\alpha_5 = \text{fix}$	Fixed tasks can only be assigned to a particular station
$\alpha_5 = \text{excl}$	Tasks may not be assigned to a particular station
$\alpha_5 = \text{type}$	Task have to be assigned to a certain type of station
$\alpha_5 = \text{min}$	Minimum distances between tasks have to be observed
$\alpha_5 = \text{max}$	Maximum distances between tasks gave to be observed
$\alpha_5 = \circ$	No assignment restrictions are considered
Processing alternatives	
$\alpha_6 = pa^{\lambda}$	Processing alternatives; with $\lambda \in \{\circ, \text{prec}, \text{subgraph}\}$
	$\lambda = \circ$: Processing times and costs are altered
	$\lambda = \text{prec}$: Precedence constraints are additionally altered
	$\lambda = \text{subgraph}$: Subgraphs are additionally altered
$\alpha_6 = \circ$	Processing alternatives are not considered

Table 3.3 The classification scheme of Boysen et al. [2007] for the stations and line characteristics

Station and Line Characteristics	
Movement of workpieces	
$\beta_1 = o\lambda\upsilon$	Paced line; with $\lambda \in \{o, each, prob\}$ and $\upsilon \in \{o, div\}$
	$\lambda = o$: (Average) work content restricted by cycle time
	$\lambda = each$: Each model must fulfill the cycle time
	$\lambda = prob$: Cycle time is obeyed with a given probability
	$\upsilon = o$: Single global cycle time
	$\upsilon = div$: Local cycle times
$\beta_1 = unpac^\lambda$	Unpaced line; with $\lambda \in \{o, sync\}$
	$\lambda = o$: Asynchronous line
	$\lambda = syn$: Synchronous line
Line layout	
$\beta_2 = o$	Serial line
$\beta_2 = u^\lambda$	U-shaped line; with $\lambda \in \{o, n\}$
	$\lambda = o$: The line forms a single U
	$\lambda = n$: Multiples Us forming an n-U line
Parallelization	
$\beta_3 = pline^\lambda$	Parallel lines
$\beta_3 = pstat^\lambda$	Parallel stations
$\beta_3 = pwork^\lambda$	Parallel working places within a station
$\beta_3 = o$	Neither type of parallelization is considered
$\lambda \in \{o, 2, 3, ...\}$: Maximum level of parallelization; o = unrestricted	
Resource assignment	
$\beta_4 = equip$	Equipment selection problem
$\beta_4 = res^\lambda$	Equipment design problem; with $\lambda \in \{o, 01, max\}$
	$\lambda = 01$: If two tasks share a resource, investment costs are reduced at a station
	$\lambda = max$: Most challenging task defines the needed qualification level of a resource
	$\lambda = o$: Other type of synergy and/or dependency
$\beta_4 = o$	Resources are not considered explicitly
Station-dependent time increments	
$\beta_5 = \Delta t_{unp}$	Unproductive activities at a station are considered
$\beta_5 = o$	Station-dependent time increments are not regarded
Additional configuration aspects	
$\beta_6 = buffer$	Buffers have to be allocated and dimensioned
$\beta_6 = feeder$	Feeder lines are to be balanced simultaneously
$\beta_6 = mat$	Material boxes need to be positioned and dimensioned
$\beta_6 = change$	Machines for position changes of workpieces required
$\beta_6 = o$	No additional aspects of line configuration are regarded

Table 3.4 The classification scheme of Boysen et al. [2007] for the objectives

Objectives	
$\gamma = m$	Minimize the number of stations m
$\gamma = c$	Minimize the cycle time c
$\gamma = E$	Maximize line efficiency E
$\gamma = Co$	Cost minimization
$\gamma = Pr$	Profit maximization
$\gamma = SSL^{\lambda}$	Station times are to be smoothed; with $\lambda \in \{stat, line\}$ λ = stat: Within a station (horizontal balancing) λ = line: Between stations (vertical balancing)
$\gamma = score$	Minimize or maximize some composite score
$\gamma = o$	Only feasible solutions are searched for

As Boysen et al. [2007] already predict and recommend, the classification may be extended to include further developments or problem variations. When needed, the new items are explained in the next sections as they appear. Along with the presented classification scheme based on $\{\alpha, \beta, \gamma\}$, the source and type of uncertainty are described, as well as the modeling approach concerning restrictions and objective function.

3.4   Uncertainties in Single-model Production Systems

3.4.1   Uncertainties in the Balancing

Since 2007, different approaches have been published, so that the $\{\alpha, \beta, \gamma\}$ classification of Boysen et al. [2007] has to be extended. Table 3.5 contains the new elements used in the classification of balancing papers with uncertainties in the processing times. The most considerable differences are different objective functions and the further division of the uncertainty factors. Uncertain processing times, for instance, are divided into stochastic (*sto*), within intervals (*int*), fuzzy (*fuzzy*), worker dependent (*worker*), and compressible (*comp*).

Table 3.5 Extra elements to the classification Boysen et al. [2007] for the single model contributions

Precedence Graph Characteristics	
Processing times	
$\alpha_3 = t^{int}$	Processing times within intervals $[a, b]$
$\alpha_3 = t^{fuzzy}$	Fuzzy processing times
$\alpha_3 = t^{worker}$	Processing time dependent on worker
$\alpha_3 = t^{comp}$	Compressible processing times
Worker requirements	
$\alpha_7 = mWO^\lambda$	Multiple (λ) workers are simultaneously required for the task
$\alpha_7 = o$	Tasks are performed by only one worker or machine
Product demand	
$\alpha_8 = dem^{sce}$	The product demand is modeled as a collection of possible scenarios
$\alpha_8 = o$	Demand is deterministic and known
Station and Line Characteristics	
$\beta_1 = prob^\theta$	$\theta = 2$: Cycle time is obeyed with a given probability in both sides of a two-sided station
	$\theta = u$: Cycle time is obeyed whit a given probability in both legs of a station in a u-line
	$\theta = o$: Cycle time is obeyed whit a given probability in a station of a straight line
$\beta_1 = \prod prob$	Cycle time is obeyed with a given probability in all stations simultaneously
$\beta_1 = \Gamma$ tasks	Cycle time is obeyed even with the deviation of the processing times of Γ tasks in a station
Objectives	
$\gamma = m^2$	Minimization of the number of stations and length of the line of a two-sided assembly line
$\gamma = c^{WC}$	Minimization of worst-case scenario (WS) for the cycle time
$\gamma = idle$	Minimization of the sum of the idle time
$\gamma = min\ prob$	Maximization of the minimal probability of non completion in any station
$\gamma = \sum prob$	Minimization of the sum non completion probability of the stations
$\gamma = RB$	Minimization/maximization of a rebalancing measure
$\gamma = Co^\lambda$	$\lambda = m$: Station or work cost minimization
	$\lambda = pstat$: Duplication cost minimization
	$\lambda = \prod prob$: Minimization of the cost of processing failure or scrappage because of incompletion
	$\lambda = compPT$: Minimization of the cost compressible processing times
	$\lambda = inc$: Minimization of the incompletion cost
$\gamma = E(UW)$	Minimization of the expected utility work
Subscript $_f$	Fuzzy objective

Table 3.6 Abbreviations used for the papers' contributions

Contributions	
BB	Branch-and-bound
BD	Benders' decomposition
CA	Complexity analysis
ChC	Chance-constraint formulation
DP	Dynamic programming
DW	Dantzig-Wolfe decomposition
GA	Genetic algorithm
GR	Graph-based approach
HI	Improvement heuristic
HS	Heuristic approach
M	Mathematical programming model
OMH	Other metaheuristic approach
PSO	Particle swarm optimization
SA	Simulated annealing
StA	Stability or sensitivity analysis

The extension of the classification shows the richness of possibilities to model stochasticity in assembly lines. The approaches differ based on the modeled assembly line, the reactions to the stochasticity (as shown in Figure 3.1 on page 33), and whether the effect of stochasticity is modeled as a restriction or aimed for at the objective function. The list of abbreviations used for the contributions is given in Table 3.6.

The proposed classification divides the literature into three groups: stochastic approaches which do not consider remedial actions; stochastic approaches which directly consider the cost of remedial actions; and robust optimization approaches.

Stochastic approaches that do not consider remedial actions

One of the simplest formulations for balancing with stochastic processing times considers that every assignment has to be feasible with respect to the cycle time for a given probability. This approach is already present in the classification of Boysen et al. [2007] as simply $\beta_1 = \text{prob}$ and has its first reference on the heuristic works by Kao [1976, 1979]. In this approach, the reliability levels used are fixed at 90% or 95%. Other solution procedures for this problem definition, including exact approaches, are given by Betts and Mahmoud [1989]; Carraway [1989]; Henig

[1986]; Leitold et al. [2019]; Nkasu and Leung [1995]. Sphicas and Silverman [1976] show that this problem is equivalent to the deterministic approach for processing time distributions with only one parameter (Poisson, gamma, binomial, negative binomial, chi-square, and normal distribution with the processing time variation as a fixed proportion of the average processing time $\sigma^2 = k \cdot \mu$). The probability constraint is also extended to other forms of assembly lines. U-shaped lines present stations that contain tasks in both the beginning and the end parts of the stations. The probabilistic constraint is then adapted to consider both parts of the station [Ağpak and Gökçen 2007; Bagher et al. 2011; Chiang and Urban 2006; Guerriero and Miltenburg 2003; Urban and Chiang 2006]. Similarly, two-sided assembly lines have mated stations, which must be considered together for the probability of completion [Özcan 2010; Chiang et al. 2016]. Delice et al. [2016] consider two-sided U-shaped lines, for which both concepts must be combined.

Stochastic problems with $\beta_1 = $ prob are easily integrated into other methods [Boysen and Fliedner, 2008]. For a given assignment and a confidence level α, the feasibility check is a simple calculation. This modeling option is therefore well suited to heuristic or metaheuristic approaches in further ALBP extensions [Cakir et al., 2011; Dong et al., 2018; Raouf and Tsui, 1982; Tang et al., 2017]. For exact methods, the feasibility test is also easy when the enumeration of the solution is performed station-wise [Henig, 1986]. For task-based enumerations, dominance rules that are valid for the SALBP version of the problem cannot be used or must be adapted to the stochastic problem [Carraway, 1989; Kao 1979, 1976; Sniedovich 1981]. The probabilistic constraints for this class of problem are also integrated into some variants of mathematical programming. Ağpak and Gökçen [2007]; Özcan [2010]; Urban and Chiang [2006]; Nourmohammadi et al. [2019] present chance-constraints which are used in mixed-integer programming or non-linear programming formulations.

A related variation of the $\beta_1 = $ prob model approach is the $\beta_1 = \prod$ prob presented by Liu et al. [2005]. The authors use a restriction on the reliability of the entire assembly line, calculated as the product of the individual stations' reliability. This measurement does not allow to consider the stations independently as $\beta_1 = $ prob so that the entire assignment must be evaluated as a whole. The interdependence of stations brings difficulties for procedures based on incomplete enumeration such as branch-and-bound or dynamic programming since dominance rules or bounds are more difficult to obtain. Liu et al. [2005], however, propose heuristic procedures for start solutions and their improvement.

Another possibility of modeling uncertainty without considering the remedial actions is based on an indirect measure or a probability minimization. Starting by indirect measures, the objective functions $\gamma = \text{SSL}^{\text{stat}}$ and $\gamma = \text{SSL}^{\text{line}}$, from the classification by Boysen et al. [2007], correspond to the horizontal and vertical balancing, respectively. These approaches are used mostly for the balancing of multiple product models, but they also find application in stochastic formulations. The first approach is due to Moodie and Young [1965]; Reeve and Thomas [1973], who assume that a well-distributed workload will result in a high-quality balancing. This objective is also used in a fuzzy approach by Tsujimura et al. [1995] and in one objective of several multi-objective approaches [Saif, Guan, Wang and Mirza, 2014; Saif, Guan, Liu, Zhang and Wang, 2014; Suresh and Sahu, 1994; Suresh et al., 1996; Zacharia and Nearchou, 2012]. The direct minimization of the non-completion probabilities may be present in different forms. Henig [1986] originally proposes a dynamic programming approach to minimize the number of stations subject to a minimal completion probability. The author shows that the enumeration procedure can be altered to maximize the stations' reliability for a given cycle time. In this approach, the minimal level of any station reliability is then maximized ($\gamma = \min$ prob). The sum of the individual completion probabilities ($\gamma = \sum$ prob) appears in compound objective functions as in Baykasoğlu and Özbakır [2007]; Saif et al. [2017]; Shin and Min [1991] or as one of multiple objectives in Saif, Guan, Wang and Mirza [2014]. Similarly, the product of the completion probabilities ($\gamma = \prod$ prob) appears in Saif et al. [2017]; Suresh and Sahu [1994]; Suresh et al. [1996].

Stochastic approaches based on the cost of remedial actions
The most common remedial actions are summarized in Figure 3.1 on page 33 and include the use of utility work, line stoppage, scrappage or disposal of the product, or repair at the end of the line. Depending on the used method, the operational cost and the respective solution method for the problem changes. Therefore, cost minimization approaches are strongly linked to the physical implementation of a system.

The active reactions with compensations within a station assume that the uncertainty will be dealt with before the product reaches the next station. This assumption simplifies an optimization procedure since it allows to treat stations independently. That is, the incompletion of one station does not affect its successors. Although this method is simpler than the within-line compensation, only a few approaches explicitly minimize the within-station compensation costs. The only example encountered is given by Shin and Min [1991], for which the corrections are made by neighboring workers, managers, or utility workers. The authors minimize the cost based on the sum of the probabilities of when workers need aid ($\gamma = \mathrm{Co}^{\sum prob}$). This objective function is related to the number of times a utility worker is required, which multiplied by a cost factor is an expectation of the utility-work cost. An alternative would be to compute the required utility work duration to estimate this cost.

A second remedial action is strongly related to the product of the completion probabilities ($\gamma = \prod$ prob). Carter and Silverman [1984] minimize the cost under the assumption that the non-completed products are discarded. As there is a probability of non-completion in each station, the overall completion probability is based on the product of the individual station probabilities.

The cost of stopping the assembly line ($\gamma = \text{Co}^{stop}$) if any station requires more processing time than the cycle time is explored by Lyu [1997] as well as Silverman and Carter [1986]. The related cost is also correlated to the product of the individual completion probabilities, but they should be weighted by the stoppage duration. Since the cost computation depends on the numerical calculation of an integral, only heuristic approaches are used for this class of problem.

Accounting for incompletion costs which are finished at the end of the line is the modeling choice for most of the cost-oriented literature on stochastic processing times. The first paper for this approach is due to Kottas and Lau [1973], who only describe an approximation for the problem. In the case of incompletion, not only the non-finished task has to be performed again at the end of the line, but all its successors are also deemed to be incomplete. In their first approach, Kottas and Lau [1973] consider the incompletion probabilities for each station independently, only summing the cost of reworking the task and their successors. In two follow-up papers, Kottas and Lau [1976, 1981] also consider the interrelationships between stations, since a non-completed task alters the expected processing times of the next stations and their respective completion probability. For this class of problem, not only the assignment of tasks to stations is important, but also their sequence within a station. As tasks are performed in a given order, the last task has a higher non-completion probability. The problem is also considered by Sarin and Erel [1990; Sarin et al. [1999; Shin and Min [1991] in its base form, Erel et al. [2005] in U-lines, Gamberini et al. [2009, 2006] in the rebalancing context, and Shtub [1984] for parallel workers. The only non-heuristic approach is a dynamic program by Sarin and Erel [1990], which is limited to small instances. Larger instances are solved with a heuristic version of the enumeration procedure. A simplification of the problem is found in Shin and Min [1991], who consider that a non-finished product causes empty cycles in the following stations, instead of allowing the non-blocked tasks to be performed.

Approaches based on solution robustness

The robustness of a solution already appeared as a research topic in terms of processing-time uncertainty, data uncertainty, and system reliability.

An important trend of approaches based on solution robustness for the balancing of assembly lines under uncertain processing times is based on processing-time intervals. An interval $[a, b]$ reflects that the lower bound of the distribution a is the normal processing time of a task. It is, however, possible due to the uncertainties, that the processing time increases up to a value b. These robust optimization approaches are based upon the supposition that not all tasks would assume their upper value at the same time.

The first mention of robustness on a balancing problem is due to Sotskov et al. [2006], who propose a stability analysis of balancing solutions. They define the stability radius of a solution as the amount of processing time that can be changed without invalidating the solution feasibility and optimality for the minimization of the number of stations. A task assignment with the sum of processing times equal to the cycle time, for instance, has a stability radius of zero, since any increase will make the assignment infeasible. Sotskov et al. [2006] propose an algorithm for the calculation of the radius for a given balancing solution. Further studies explore the stability of solutions for other objective functions, such as cycle time [Lai et al., 2016], line efficiency [Gurevsky et al., 2012; Lai et al., 2019], and number of stations under the possibility of parallelization [Gurevsky, Battaïa and Dolgui 2013].

Dolgui and Kovalev [2012] conduct a complexity analysis of assembly-line problems under uncertain processing times within intervals. They consider several variations of the problem and prove that the balancing problem for cycle time minimization under scenarios of processing times is strongly NP-Hard. Furthermore, the authors propose dynamic programming approaches for some classes of the problem.

The robust approach can also be defined as an optimization problem. For this class, the restriction $\beta_1 = \Gamma$ tasks is usually applied. In this approach, robustness is defined as the ability to assure feasibility even if Γ tasks (and alternatively a given percentage of tasks) assume their maximal value of the interval processing time. This problem is solved for different objective functions, such as the minimization of the number of stations [Gurevsky et al., 2012; Pereira and Álvarez-Miranda 2018], the minimization of the cycle time [Hazir et al., 2013] or for different line characteristics such as U-shaped lines [Hazir et al., 2015] or heterogeneous workers [Moreira et al., 2015]. An optimization of a different perspective is given by Rossi et al. [2016], who do not search for the best cycle time or the number of stations, but for the most stable solution in terms of stability radius.

A second modeling possibility for the robust optimization problem is to consider either the worst case or the α-worst case scenario. The α-worst case scenario optimizes for the α percentile solution. That is, for a value of 90%, the worst 10% are not considered in the computation. This way, an assignment is robust for most of the scenarios, but it does not need to be over-pessimistic in considering all possible scenarios. The α-worst robust optimization for stochastic processing times can be found in Saif, Guan, Liu, Zhang and Wang [2014]. Pereira [2018] considers the combination of all possible scenarios for the processing time based on intervals. His procedure minimizes the *regret* of the worst-case scenario. Regret is defined as the difference between the objective value of the problem and the optimal value for the specific scenario. The formulation uses a min-max expression for the regret based on the cycle time using heterogeneous workers.

An alternative approach to consider robustness is fuzzy modeling. In this approach, the data is considered uncertain and is described as linguistic variables. In the assembly line balancing context, processing time, cycle time, and even the number of stations can be considered fuzzy. Tsujimura [1995] present the first contribution in the balancing literature, in which fuzzy processing times are smoothened among the stations. Other versions of the problem, such as the minimization of the number of stations (along with a second smoothening objective) is presented by Zacharia and Nearchou [2012] and the maximization of line efficiency by Zacharia and Nearchou [2013].

In a different field of research, Müller et al. [2016, 2018] solve the balancing problem with the objective of increasing the robustness of the line with respect to failures. They consider that one station can serve as a backup for other stations if the stations are equipped with compatible equipment and the precedence constraints are obeyed. In this formulation, if a failure occurs in any station, a containment plan is already prepared and optimized to redistribute the tasks between the remaining stations. As this approach is different from all the others in the literature, the characteristics of machine failure and redundancy maximization are not added to the classification scheme.

Summary of the literature classification
The summary with all reviewed references for the stochastic balancing of single-model assembly lines with uncertainty in the processing time is given in Table 3.7.

Some terms in the table are not contained in the classification: Müller et al. [2016, 2018] maximize the *redundancy* of an assembly line considering machine *failures*; the *Part-feeding* objective from Nourmohammadi et al. [2019] refers to a logistic objective function and is therefore not added to the classification scheme; Rossi et al. [2016] maximize the stability radius of an optimal solution in terms of

Table 3.7 Literature overview for the stochastic balancing of single-model assembly lines

Author	Characteristics	Contribution
Ağpak and Gökçen [2007]	[t^{sto} | prob^u, u | m]	M, HS, ChC
Bagher et al. [2011]	[t^{sto} | prob^u, u | m]	OMH
Baykasoğlu and Özbakır [2007]	[t^{sto} | u | m + idle + $\sum$ prob]	GA
Betts and Mahmoud [1989]	[t^{sto} | prob | m]	HS
Boysen and Fliedner [2008]	[t^{sto}, link, inc, cum, pa | prob, pstat, ptask, res01, resMax, u | Pr]	HS, GR
Cakir et al. [2011]	[t^{sto} | prob, pstat | SSL^{stat}; m^2]	SA
Carraway [1989]	[t^{sto} | prob | m]	DP
Carter and Silverman [1984]	[t^{sto} | | Co^{m}, $\prod^{prob}$]	HS
Chiang and Urban [2006]	[t^{sto} | prob^u, u | m]	HS, HI
Chiang et al. [2016]	[t^{sto} | pwork^2, prob^2 | m^2]	PSO
Delice et al. [2016]	[t^{sto} | pwork^2, prob^{2,u}, u | m^2]	GA
Dolgui and Kovalev [2012]	[t^{int} | | c]	CA
Dong et al. [2018]	[t^{int}, t^{comp}, link, exc | prob | c; Co^{compPT}]	PSO, SA
Erel et al. [2005]	[t^{sto} | u | Co$^{m;inc}$]	HS
Gamberini et al. [2006]	[t^{sto} | | Co$^{E;inc}$; RB]	HS
Gamberini et al. [2009]	[t^{sto} | | Co$^{E;inc}$; RB]	HS, GA
Guerriero and Miltenburg [2003]	[t^{sto} | prob^u, u | m]	GR
Gurevsky et al. [2012]	[t^{int} | | E]	StA
Gurevsky, Hazir, Battaïa and Dolgui [2013]	[t^{int}, exc | Γ tasks | m]	BB
Gurevsky, Battaïa and Dolgui [2013]	[t^{int}, exc | pstat | Co$^{m;pstat}$]	StA
Hazir et al. [2013]	[t^{int} | Γ tasks | c]	BD
Hazir et al. [2015]	[t^{int} | Γ tasks, u | c]	BD
Henig [1986]	[t^{sto} | prob | m / c / min prob]	DP
Kao [1976]	[t^{sto} | prob | m]	DP
Kao [1979]	[t^{sto} | prob | m]	DP
Kottas and Lau [1973]	[t^{sto} | | Co^{inc}]	HS

(continued)

Table 3.7 (continued)

Author	Characteristics	Contribution
Kottas and Lau [1976]	[t^{sto} | | Co^{inc}]	HS
Kottas and Lau [1981]	[t^{sto} | | Co^{inc}]	HS
Lai et al. [2016]	[t^{int}, exc | | c]	StA
Lai et al. [2019]	[t^{int}, exc | | E]	StA
Leitold et al. [2019]	[t^{sto} | prob | m]	SA
Liu et al. [2005]	[t^{sto} | $\prod$ prob | c]	HS, HI
Lyu [1997]	[t^{sto} | | Co^{stop}]	HS
Moodie and Young [1965]	[t^{sto} | | SSL^{line}]	HS
Moreira et al. [2015]	[t^{int}, t^{Worker} | Γ tasks | m]	M, HS
Müller et al. [2018]	[| equip, failure | redundancy]	M
Müller et al. [2018]	[| equip, failure | redundancy]	GA
Nkasu and Leung [1995]	[t^{sto} | prob | m / c / E]	HS
Nourmohammadi et al. [2019]	[t^{sto}, Task Demand^{sto} | prob, u | m, Part-feeding]	M, HS, ChC
Özcan [2010]	[t^{sto} | prob^2, pwork^2 | m^2]	M, SA, ChC
Pereira and Álvarez-Miranda [2018]	[t^{int} | Γ tasks | m]	BB, DW, HS
Pereira [2018]	[t^{int}, t^{worker} | | c^{WC}]	M, HS, BB
Raouf and Tsui [1982]	[t^{sto}, fix, exc | prob | m / SSL^{stat}]	HS
Reeve and Thomas [1973]	[t^{sto} | | SSL^{line}]	HI, BB
Rossi et al. [2016]	[t^{int} | | Stability radio]	M
Saif, Guan, Liu, Zhang and Wang [2014]	[t^{sto} | | c; $\sum$ prob; SSL^{line}]	OMH
Saif, Guan, Wang and Mirza [2014]	[t^{sto} | | c^{WC}; SSL^{line}]	M
Saif et al. [2017]	[t^{sto} | | c; $\sum$ prob + $\prod$ prob]	OMH
Sarin and Erel [1990]	[t^{sto} | | Co^{inc}]	DP, HS
Sarin et al. [1999]	[t^{sto} | | Co^{inc}]	HS, HI
Shin [1990]	[t^{sto} | | Co^{inc}]	HS
Shin and Min [1991]	[t^{sto} | | $Co^{m; \sum prob}$]	HS
Shtub [1984]	[t^{sto} | pstat | Co^{inc}]	M, HS
Silverman and Carter [1986]	[t^{sto} | | Co^{stop}]	HS

(continued)

Table 3.7 (continued)

Author	Characteristics	Contribution
Sniedovich [1981]	[t^{sto} | prob | m]	DP
Sotskov et al. [2006]	[t^{int} || m]	StA
Sphicas and Silverman [1976]	[t^{sto} | prob | m]	M
Suresh and Sahu [1994]	[t^{sto} || SSL^{line} / $\prod$ prob]	SA
Suresh et al. [1996]	[t^{sto} || SSL^{line} / $\prod$ prob]	GA
Tang et al. [2017]	[t^{sto}, link, exc, mWo^2 | pwork^2, prob^2 | m^2]	OMH
Tsujimura et al. [1995]	[t^{Fuzzy} || SSL^{line}]	M, GA
Urban and Chiang [2006]	[t^{sto} | prob^u, u | m]	M, ChC
Zacharia and Nearchou [2012]	[t^{fuzzy} || Score$_f^{m,\text{SSL}^{\text{line}}}$]	GA
Zacharia and Nearchou [2013]	[t^{fuzzy} || E$_f$]	GA
Zhang et al. [2014]	[t^{sto} || CT; Cost(?)]	OMH
Zhang et al. [2017]	[t^{sto} || CT; Cost(?)]	OMH

the number of tasks with possibly increased processing times; Zhang et al. [2014, 2017] minimize the assignment costs based on scenarios, although it remains to be specified, what these costs represent. Objective functions separated by a semi column indicate multiple objective functions, the slash (/) symbol is used for alternative objective functions, while the plus sign (+) indicates that the objective function is a (weighted) sum of multiple terms. The abbreviations used in Table 3.7 are explained in Table 3.6.

3.4.2 Uncertainties in the Buffer Allocation

The Buffer Allocation Problem (BAP) is considered a stochastic problem for all the references of the surveys presented in Section 3.2.6. Weiss et al. [2019] present a very extensive, detailed, and up-to-date review on the BAP and its variations. As previously described in Section 3.2.6, the buffer allocation problem usually assumes a production system to be given, while the buffer allocation is the only decision variable. A handful of references extends this scope and includes some element in the production system as a variable problem. Weiss et al. [2019] do not include these articles in the classification, but cite some of them in their discussion section. These and related references are discussed in the following paragraphs.

Spinellis et al. [2000] integrate the buffer allocation problem with the server allocation and the workload allocation problem. In the server allocation problem, the number of machines (or servers) in each step of the production line can be optimized. This results in parallel machines. Furthermore, Spinellis et al. [2000] consider that the workload can be allocated freely between the stations. That is, the expected processing time for the whole process is constant, but the amount of work in each station can be selected by the planner. With this degree of freedom, Spinellis et al. [2000] arrive at optimal solutions that contain unbalanced processing times.

An optimization software architecture is proposed by Spinellis and Papadopoulos [2001]. Their problem definition is similar to the one of Spinellis et al. [2000], although the software tool should employ a wide range of algorithms for the solution of instances.

Hillier and Hillier [2006] also consider that the total processing times of workpieces can be continuously distributed among the workstations. They perform a simultaneous optimization of the work and buffer allocation for the production with stochastic processing times. The processing-time mean at each station and the buffer sizes are the problem variables. The objective of their modeling is the maximization of the revenue minus the associated cost of the required buffer space.

Hillier [2013] extends the work of Hillier and Hillier [2006] by considering the work-in-process cost. The objective function is changed to the revenue minus the inventory costs. This objective function takes into account the buffer utilization and minimizes the monetary cost of the work-in-process contained in the buffers.

The combined buffer allocation and equipment selection is explored in Nourelfath et al. [2005]. In this work, several machines are at the disposal of the line planner. They perform the tasks at the same deterministic rate, but present different costs, failure rates, and time-to-repair rates. In Nourelfath et al. [2005], serial lines are considered in the problem definition. Nahas et al. [2009] extend the approach and consider serial or parallel machines, while a follow-up work [Nahas et al., 2014] investigates more complex networks.

Table 3.8 contains the summary of the surveyed references. The scope of the problem, its objective function, reliability, random distributions, and solution methods are described. Some abbreviations are used for the table, such as THR for throughput maximization, BUF for buffer cost, and WIP for work-in-progress cost. The processing times are either deterministic (DET) or follow the exponential (EXP) the Erlang (ERL) distribution. The evaluation methods are either based on decompositions (DEC) or exact approaches using a Markov Chain (MC). Finally, the generation methods are either enumerations (ENU), heuristics ($HEUR$), or meta-

Table 3.8 Literature overview for the buffer allocation problem with variable production processing times. OF stands for Objective Function; PTD stands for Processing Time Distribution; FRR stands for Failure and Repair Rate distribution; Eval. and Gen. stand for the evaluation and generation methods, respectively; and Size is measured in the number of stations

Author(s) (Year)	Problem	Layout	OF	PTD	FRR	Eval.	Gen.	Size
Spinellis et al. [2000]	Workload allocation	Parallel	THR	EXP	–	DEC	SA	60
	Server allocation							
Spinellis et al. [2005]	Equipment selection	Serial	THR	DET	EXP	DEC	ACO	10
Hillier and Hillier [2006]	Workload allocation	Serial	THR – BUF	EXP /	–	MC	ENU	5
				ERL				
Nahas et al. [2009]	Equipment selection	Parallel	THR	DET	EXP	DEC	SA /	20
							ACO	
Hillier [2013]	Workload allocation	Serial	THR – WIP	EXP /	–	MC	ENU /	4
				ERL			HEUR	7
Nahas et al. [2014]	Equipment selection	Network	THR	DET	EXP	DEC	GA	15

heuristics (Simulated annealing - SA, ant colony optimization - ACO, or genetic algorithm - GA).

Table 3.8 contains only the references with buffer allocation problems integrated with other problems. For the BAP works without the integration see the extensive review of Weiss et al. [2019]. All the displayed references and the majority of the works cited by Weiss et al. [2019] contain exponential, or Erlang distributed processing times. This assumption is not usually encountered in assembly lines of the automotive industry. Furthermore, the size of buffers for products of the size of vehicles limits their use in real assembly lines.

3.5   Uncertainty in Multiple-model Production Systems

The processing times in stations of an assembly line producing multiple models depend on more factors than the case of the single-model variant. Besides the possible uncertainty of the processing times of each model, the characteristics of the product itself may be uncertain. This uncertainty is due to the multiple products which may be assembled within the same line. More specifically, the production demand or the production sequence can be modeled as random variables and may account for the uncertainty in multiple-model production systems.

In this section, the literature is divided into three subsets, depending on the source of uncertainty. The first group is similar to the single-model case and considers uncertain processing times (t^{sto}, t^{int}, or t^{fuzzy}). The second group contains the literature on assembly lines with a given and deterministic demand, which is produced in a random order (rnd seq). That is, the production sequence is randomly determined based on random draws of the products' relative demands. The third group contains the literature on demand uncertainty. For this case, the relative demands themselves are considered to be random variables (dem^{sce}). The summary of all articles is given in Table 3.9 on page 44.

A second classification is related to how the multiple models are handled in the solution procedure. Five approaches are identified and are included in the last column of the literature review of Table 3.9. The simplest treatment is to reduce the multiple models to an equivalent average single model ($mix - single$). A second common approach corresponds to the *each* classification of Table 3.3, for which all restrictions are applied to all models individually. However, restricting all models equally may be too restrictive as the whole line can then be sized for the most loaded model. The third approach (div) also treats each model individually but uses different levels of restriction for each model. The fourth and fifth formulations deal with the interaction of the different models in their production sequence. The acronym *seq* is used for the procedures which integrate the optimization problem of sequencing the production along with the balancing of the assembly line. Another form to integrate explicitly the production order without solving for the best sequencing is to consider a random sequence ($rnd - seq$). The random sequence simulates a Just-in-Sequence production [Bukchin et al., 2002], in which the products are assembled in the sequence they are ordered by the end customer. These orders are modeled as a random sequence based on the relative demand of the products.

3.5.1   Uncertainty in the Processing Time

Among the literature on multiple models with uncertain processing times, most of the papers deal with the models either separately or aggregate them into an average model ($mix - single$). Vrat and Virani [1976] reduce a mixed-model assembly line into a single model line with stochastic processing times. The studies by McMullen and Frazier [1997]; McMullen et al. [1998]; McMullen and Tarasewich [2003, 2006] consider several objectives. For some of them, the average model is used.

The application of the restrictions to all models individually is present in Hop [2006], who model a mixed-model assembly line with fuzzy processing times. The div category is represented by Chakravarty and Shtub [1986] and Al-E-Hashem et al. [2009]. Chakravarty and Shtub [1986] integrate the balancing problem with the lot-sizing of multi models so that the production and the cycle times of each model are taken into account independently. Al-E-Hashem et al. [2009] model the robust assembly-line balancing of a mixed-model assembly line, in which each model has its own cycle time.

The combined balancing and sequencing problem of mixed-model assembly lines under uncertainty is represented by Özcan et al. [2011], Dong et al. [2014], and Tiacci and Mimmi [2018], classified within the *seq* criterion. Özcan et al. [2011] deal with the balancing and sequencing of U-shaped mixed-model assembly lines. As stations of U-shaped lines may produce different models on each of the sides of the station, the processing time of the station depends on the sequence of products. Therefore, Özcan et al. [2011] perform a vertical balancing considering the sequence of the products. Dong et al. [2014] solve the combined balancing and sequencing problem minimizing the expected utility work for paced lines. For unpaced lines, Tiacci and Mimmi [2018] propose a genetic algorithm that solves the balancing and sequencing problem simultaneously.

3.5.2 Uncertainty of the Production Sequence

The production of a random production order is called by Bukchin et al. [2002] a Just-in-Sequence production, in which the end customer orders are produced in the sequence they are received. In the first work on this setting (*rnd − seq*), Bukchin et al. [2002] propose a heuristic procedure that maximizes an approximation of the throughput. The objective function is based on the *bottleneck measure* that estimates the cycle time of unpaced assembly lines under random sequences. Manavizadeh et al. [2012] extend the work of Bukchin et al. [2002] with respect to multiple objective functions.

Another approach to model random sequences (*rnd − seq*) relies on simulation. Tiacci [2015a] and Tiacci [2015b] propose two genetic algorithms for the throughput maximization of unpaced assembly lines under random sequences and stochastic processing times. Tiacci [2015b] also integrates the buffer allocation into the problem.

3.5.3 Uncertainty of Demand

The third source of uncertainty lays in the demand for multiple products. These uncertainties are not explored very much in the literature, although they are often observed in practice [Chica et al., 2016]. The planning of a new assembly line is performed much earlier than the production phase. Between the planning and the realization, a planner must rely on demand estimates. Furthermore, the demand may change throughout the lifetime of an assembly line. Not only do the weekly or monthly sales vary, but also new models are developed while others are taken out of the market.

All of the references on demand uncertainty model demand as given in a finite number of possible scenarios. Simaria et al. [2009] model a multi-model assembly line, in which only one model can be produced at a given time. In their problem, the demand for each product is given in terms of specific cycle times. For the different scenarios, the number of workers can be adjusted to the demand. That is, the assignments of the tasks are fixed, but the workers can be reassigned to properly match the demand scenarios. Another example of different demand level scenarios is given by Li and Gao [2014]. In their approach, overtime is used to compensate for the most loaded demand scenarios. Li and Gao [2014] model a mixed-model assembly line as its average single model equivalent and minimize the total cost comprising of normal and overtime wages.

Other authors explore the demand uncertainty in the context of robust optimization: Chica et al. [2016] and Chica et al. [2019], for example, consider the robust assembly-line balancing problem with space limitations. In their problem, tasks need space for equipment and for the inventory of the parts that are mounted into the product. Therefore, not only the processing time must be accounted for, but also the needed space of each task. They deal with the multi-objective optimization of such lines, in which the robustness in terms of the number of stations, cycle time, and the area is accounted for. To model the multiple models, the average of the processing time and area requirement is used for each demand scenario. For the multiple objective functions proposed, either the number of non-achieved requirements or the maximal amount of time and/or space over the capacity is minimized. A different approach is due to almost identical papers by Xu and Xiao [2009] and Xu and Xiao [2011]. They consider mixed-model assembly lines with stochastic processing times and demand based on scenarios. The objective function is the minimization of the variance between the stations' workloads (vertical balancing). They consider the worst case and the α worst case (α percentile) of the variation for all possible scenarios. However, it remains to be clarified how the multiple models are dealt with besides the calculation of the processing-time variance. No feasibility condition based on the individual models or the average model is discussed, so that the column 'Models' from Table 3.9 is left blank for these two papers.

Table 3.9 Literature overview for the stochastic balancing of multiple-model assembly lines

Author	Characteristics	Contribution	Models
Al-E-Hashem et al. [2009]	[mix, t^{int} | div, equip, $\Gamma task$ | $Co^{m,Eq}$]	M	div
Bukchin et al. [2002]	[mix | unpac | c]	HS, HI	rnd seq
Chakravarty and Shtub [1986]	[mult, t^{sto} | div, Δt_{unp}, buffer | $Co^{m,stor,set-up}$]	HS	div
Chica et al. [2016]	[mix, cum, dem^{sce} | each | m, c, area, robust]	EA	mix-single
Chica et al. [2019]	[mix, cum, dem^{sce} | each | m, c, area, robust]	EA + Sim	mix-single
Dong et al. [2014]	[mix, t^{sto} | u | E(UW)]	SA	seq
Hop [2006]	[mix, t^{fuzzy} | each | m]	M, HS, HI	each
Li and Gao [2014]	[mix, dem^{sce} | | $Co^{m,c}$]	HS, BBR	mix-single
Manavizadeh et al. [2012]	[mix | | m, c, SSL^{line}, SSL^{stat}]	EA	rnd seq
McMullen and Frazier [1997]	[mix, t^{sto} | pstat | $Score^{m;pstat,SSL^{line},\prod prob}$]	HS	mix-single
McMullen et al. [1998]	[mix, t^{sto} | pstat | $Score^{m;pstat,SSL^{line},\prod prob}$]	SA	mix-single
McMullen and Tarasewich [2003]	[mix, t^{sto} | pstat | $Score^{m;pstat,SSL^{line},\prod prob}$]	ANT	mix-single
McMullen and Tarasewich [2006]	[mix, t^{sto} | pstat | $Co^{(m;pstat)}$, SSL^{line}, $\prod$ prob]	ANT	mix-single
Özcan et al. [2011]	[mix, t^{sto} | prob, u | SSL^{line}]	GA	seq
Simaria et al. [2009]	[mult, exc, link, dem^{sce} | u, div | $Score^{E,idle}$]	ACO	div
Tiacci [2015a]	[mix, t^{sto} | pstat, equip, unpaced | $Co^{m,pstat,equip}$]	GA + Sim	rnd seq
Tiacci [2015b]	[mix, t^{sto} | pstat, equip, buffer, unpaced | $Co^{m,pstat,equip,buffer}$]	GA + Sim	rnd seq
Tiacci and Mimmi [2018]	[mix, t^{sto} | erg | pstat, unpaced, res^{01} | $Score^{m,pstat,equip,erg}$]	GA + Sim	seq
Vrat and Virani [1976]	[mix, t^{sto} | pstat | Co^{inc}]	HS	mix-single
Xu and Xiao [2009]	[mix, t^{sto}, dem^{sce} | prob | $SSL^{line}\alpha_{WC}$]	GA	
Xu and Xiao [2011]	[mix, t^{sto}, dem^{sce} | prob | $SSL^{line}\alpha_{WC}$]	GA	

Some minor terms are left out of the classification here because they only apply to one paper and are only partly related to the balancing problem. The cost of inventory storage $(stor)$ in Chakravarty and Shtub [1986], set-up costs $(set-up)$ in Chakravarty and Shtub [1986], and ergonomic factors (erg) in Tiacci and Mimmi [2018] are some examples. Furthermore, $robust$ is used as a robustness expression for the objective function as in Chica et al. [2016, 2019].

3.6   Uncertainties of the Disassembly Process

The balancing of disassembly lines is first discussed by Gungor and Gupta [2001], who explain in detail the differences between the assembly and disassembly processes. The disassembly process may be much more variable since there exists a variety of possible inputs and outputs. The products to be dismantled, for instance, may comprise of different models of products, may greatly vary in quality, or even be damaged or broken. The output of the line also depends on the product and the environmental and economic restrictions. Some parts or sub-assemblies may be reused, while other parts of the product need to be discarded. Another significant difference is the precedence relations between tasks. The precedence graph of the disassembly process is not simply the inverse production order. There are, for instance, 'OR' precedence relations [Gungor and Gupta, 2001] if an internal part can be removed from two or more sides of the product. Removing either one of the blocking parts would allow the task to be performed.

The disassembly-line balancing is a very active research field and has received multiple literature reviews. A broad survey on disassembly-line balancing is provided recently by Özceylan et al. [2019]. As the survey is published in 2019, this section uses the classification and the review of the survey to discuss the particularities of uncertainty in disassembly lines.

Özceylan et al. [2019] identify 116 references from 1999 to 2018 dealing with the balancing of disassembly lines, from which 33 studies present some sort of uncertainty. Out of these 33 references, 27 of them present stochastic or fuzzy processing times as their only uncertainty. As the uncertainty of processing times is thoroughly discussed for the balancing problem in Subsection 3.4.1, only the different sources of uncertainty are discussed in this section.

Gungor and Gupta [2001] consider a probability of tasks with defects. Failure results in not being able to perform tasks and their successors, affecting the processing times in the next stations. The proposed solution method contains a network algorithm that models assignments as nodes and arcs as the costs in the objective function. The procedure initiates without the stochastic effects, which are calculated for the shortest path in each iteration. The effects on each branch are calculated based on independent probabilities for each task.

Turowski et al. [2005] consider heterogeneous workers who can also randomly damage the pieces in their operations. They propose a heuristic that maximizes the disassembly-line profit based on the idle-time minimization and the damage levels of pieces.

Altekin and Akkan [2012] also model the possibility of task failures. The authors consider the possibility of rebalancing the line after a defect. That is, the tasks can be rearranged in the remaining stations after a failure. Their objective is to maximize the profit of incomplete disassembly. The assignment of tasks defines which pieces will be removed and which are left in the used product to discard. If the tasks can be reordered after a failure, the disassembly can be changed to remove the components that are not affected by the failure. The authors consider the probability of each failure and the optimized rebalancing for each case. In their formulation, it is assumed that only one task can fail for each product.

Paksoy et al. [2013] consider the uncertainty related to objective functions. They propose a fuzzy goal programming and a fuzzy multi-objective programming model for the minimization of the cycle time, number of stations, and soothing. The uncertainty lays in the importance of the objective function terms, which may change during the lifetime of the disassembly line.

Özceylan and Paksoy [2014] integrate disassembly with a Closed-Loop Supply Chain. That is, not only the disassembly line is modeled, but also the production, distribution, and collection of the used products are treated. The costs, demand, and return rates of used products are modeled as fuzzy parameters in a fuzzy integer programming model.

Tuncel et al. [2014] use reinforcement learning to solve a disassembly-line balancing. They model a deterministic and a stochastic environment, in which the demand is uncertain. The demand only affects one of three objective functions, which minimize the time to retrieve the demanded pieces. It remains unclear why it is important that a piece is to be removed on an earlier station since the cycle time is generally small. In the steady-state, the amount of removed demanded pieces will be equal independent of the station in which they are removed. Therefore, it is questionable whether the approach is realistic. Furthermore, the demand does not alter the number of dismounted products, the cycle time, or the number of stations. The different demand levels affect only the objective function coefficients.

3.7   Gaps and Contributions to the Literature

The objective of this chapter is to give a broad overview of uncertainties considered in the balancing of assembly lines and related problems. Based on the cited papers and the surveys on this topic, it can be observed that the literature on stochastic and fuzzy problems is plenty. The approaches dealing with uncertainties are mostly focused on single-model production, while the most common uncertainty lays in the processing times. Therefore, contributions in the less explored directions are pointed out as the focus of this manuscript.

In Chapters 4 to 6, three contributions to the literature on mixed-model assembly lines are introduced. The source of uncertainty of the contributions lays in an unknown demand in Chapter 4 and unknown production sequences in Chapters 5 and 6.

In the first contribution (Chapter 4, also described in Sikora [2021]), the demand is modeled as a collection of possible scenarios, as it is employed by the other papers on demand uncertainty. Different from the approaches which deal with the average model or with the models individually (such as the robust optimization), the chosen approach is to consider the sequence of the products explicitly. The selected assembly-line layout is a straight and paced line, in which utility work can be used to assure every product is finished within the station bounds. Each demand scenario consists of the relative demand of the ordered products, so that every scenario may require a different production sequence. The assignment of tasks and equipment to the stations after the balancing is not flexible and must be the same for all demand scenarios. As the balancing and sequencing have two different time frames, the problem is modeled as a two-stage stochastic optimization problem. The balancing is determined in the first stage with only the knowledge of the demand-scenario distribution. The sequencing can be optimized after the orders (and consequently the demand) are known so that each scenario is modeled individually on the second stage. The objective of the formulation is to find the balancing solution with the fewest expected utility work for a given cycle time and number of stations.

The contribution described in Chapter 5 assumes that the decision-maker does not influence the production sequence. As reviewed in Section 3.5.2, the uncertainty of the production sequence is mostly modeled with a random sequencing simulating the customers' orders. The proposed contribution optimizes the balancing of a paced assembly line under the assumption that the products are randomly selected with a fixed relative product probability. The objective is to minimize the expected utility work to operate the assembly line, which is calculated exactly using results from Markov chains [Gwiggner, 2020]. In the literature, a similar problem is explored by Bukchin et al. [2002]. The reference, however, deals with unpaced lines and proposes a heuristic procedure.

The third contribution (Chapter 6) is a middle ground between the total sequence control of Chapter 4 and the random sequence of Chapter 5. The production sequence is considered to enter the assembly line in a random order but may be resequenced using a buffer. This contribution focuses on the optimization of buffer use. The production is considered to be a given straight and paced assembly line, in which the products flow in the order they are inserted. The optimization problem consists of choosing the product order so that the due dates are obeyed and the utility work in the assembly line is minimized. The problem is solved at every cycle time since a product must be selected to advance into the line. As the next product that enters the buffer is unknown, the optimization occurs under uncertainty in an online optimization problem.

The three contributions aid in filling out the gap of the design of assembly lines under uncertainty of what is produced. Both the demand levels as well as the production order are tackled, considering both the possibility and impossibility of production sequencing. Furthermore, an application-oriented sequencing procedure is explored, in which a policy for the buffer usage is obtained.

Balancing Under Full Sequencing Control   4

In this chapter, a solution method for the assembly-line balancing under demand uncertainty is proposed. The project is included in Sikora [2021], which is published in the European Journal of Operational Research. This chapter has two purposes: the first one is to describe the development of the solution algorithm and the tests that are included in Sikora [2021]; the second purpose is to provide complementary content that is not included in the paper, such as some preliminary tests and results on further developments. The chapter is divided into three sections, in which the problem definition, the solution algorithm, and the results are described.

4.1   Problem Definition

In this section, the modeled production system is described, as well as its parameters, uncertainties, and the order in which the decisions are made. After the problem description, a mixed-integer programming formulation is presented.

The considered production system consists of a paced assembly line. A conveyor belt is responsible for the transport of workpieces along a series of stations. The stations are ordered in a straight line and the conveyor speed is constant and is matched to the required cycle time. The workstations have fixed boundaries which do not overlap with the ones from other stations. The length of each station is constant for all stations. The stations can be longer than the cycle time equivalent space so that variations of processing times of the products can be compensated within a station. Furthermore, utility work is used as production support, if at any time a product can not be finished within the station boundaries. The objective of the problem is to balance the assembly line while minimizing the expected amount of necessary utility work.

C. G. Stall Sikora, *Assembly-Line Balancing under Demand Uncertainty,* Gabler Theses, https://doi.org/10.1007/978-3-658-36282-9_4

The efficiency of a mixed-model assembly line and the required utility work depends on the production sequence. In the literature, one form of dealing with the sequencing of assembly lines is based on a minimal part set (MPS) e.g., Bard [1989]. Such an MPS consists of a minimal representation of the relative demand for product models. For instance, the production of 300 Model A, 200 Model B, and 100 Model C products would result in an MPS of relative demands (3,2,1). Scheduling the MPS is known as an efficient form to solve the sequencing problem instead of dealing with the complete sequence [Lopes et al., 2020b]. In the example, 6 models should have to be sequenced in the MPS instead of the 600 products in the complete sequence. The MPS approach usually implies that the production would repeat the minimal sequence to achieve the total demand levels [McCormick et al., 1989].

Figure 4.1 illustrates the effect of the processing times of different models and their sequence. In this figure, an MPS comprising of four products is represented with $P1$ to $P4$ as horizontal bars. The station is defined as the space between the left and the right vertical lines, while the dashed vertical line represents the cycle-time equivalent distance. For a given conveyor speed, a piece entering the workstation at position 0 would be in position CT after a cycle time. For convenience, the conveyor speed is set to 1 length unit per time unit, so that the conveyor-belt displacement after a cycle time is also CT length units. The length of the station is also defined based on the cycle time using a length multiplier (LM). The multiple products are displayed in the same diagram, in which the y-axis represents time. Each bar stands for the processing time of a product, defining the start and end position of the processing. For piece $P1$, for instance, the worker starts the processing at the left border and finishes his operations some time after the cycle time. At the end of the processing of product $P1$, the worker moves to the next piece and starts immediately. As product $P1$ required more than the cycle time, the second product enters the line before $P1$ is finished. This way, the start position of $P2$ is larger than zero. If the worker finishes a product before the cycle time, idle time occurs, because the next product has not yet reached the station's entry point. After a sequence of products with long processing times ($P3$ and $P4$), the final position of the worker would be outside the station borders. This is avoided by using utility workers that can be called to help with the production and avoid the station limit violations. This sequence requires utility work to avoid the violation caused by $P4$.

As the MPS is quite small in relation to the full sequence, the sequencing is performed cyclically. The given MPS is repeated until the global-production demand is achieved. In order to allow a steady and cyclical production, the start and end positions of each product must be the same in each cycle. In the example of figure 4.1, it would be impossible to repeat the MPS $P1 - P4$ starting at position 0 in the second cycle at the same position, since $P1$ of the second cycle has entered the station before $P4$ is finished. This delay between two MPS can also be corrected using utility work. If a utility worker starts the processing of $P1$ at position 0, the regular worker can resume $P1$ after $P4$ and repeat the exact same cycle as the first MPS. This way, a cyclic scheduling is obtained.

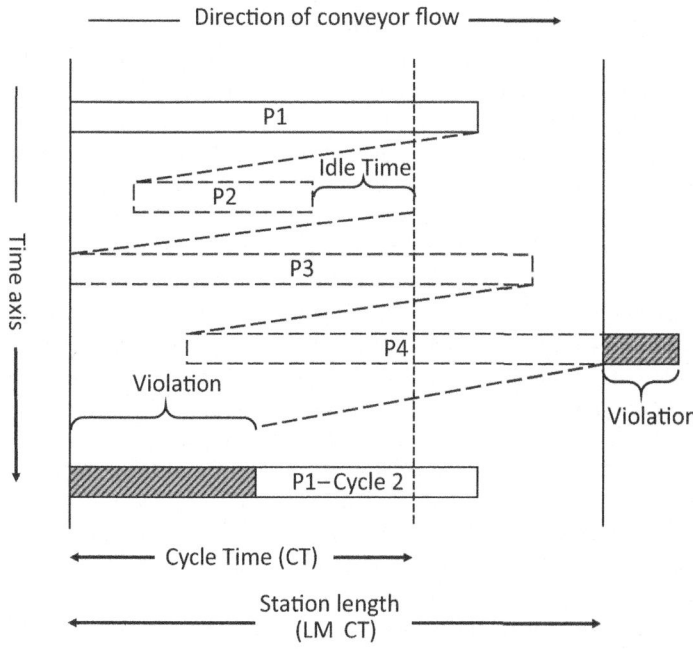

Figure 4.1 Example of the scheduling in a station of a paced assembly line. Figure from Sikora [2021]

The uncertainty of the system is derived from the product demand. This uncertainty can be interpreted as a project uncertainty, in which the assembly line has to be planed before the real demand is known. A second interpretation is the possible demand fluctuations through the lifespan of the assembly line. For this approach, the demand is defined as a collection of possible scenarios, which contains the relative demand of each product. The demand distribution is defined by a collection of MPSs and their corresponding probabilities. The minimization of the expected utility work is based on the average values of the given demand-scenario realizations.

The optimization problem in question is the design of an assembly line that operates well under uncertainty. Tasks must be assigned to workstations, setting the processing time of each produced model in each station. As multiple models exist, the operation must consider the sequence of pieces, so that both problems are interrelated. The timespan of each decision, however, greatly differs. The assignment of tasks to workstations is a planning problem that is generally performed before the building of the assembly line. As the assignments require the acquisition and installment of equipment and machinery, the balancing is considered a middle to long- term decision and cannot be adjusted easily [Boysen et al., 2009b]. The sequencing of products, on the other hand, has fewer restrictions and is considered as an operative decision [Boysen et al., 2009c].

The defined problem combines the balancing and sequencing problem in a stochastic framework. As the balancing must be performed before the real knowledge of the demand, the decision must be taken a priori. The sequencing of the products can be specified after the products are ordered so that they can be solved with full information. This structure is common in stochastic two-stage problems [Birge and Louveaux, 2011], in which the first stage must be defined with uncertainty and the second stage is solved after the realization is known.

The described problem is formulated as mixed-integer linear programming. The balancing decisions are part of the first stage and the sequencing is the second stage of a stochastic two-stage model. The model is given and described as in Sikora [2021]. This model is called "monolithic model", as all decisions (both first and second stage) are modeled in the same formulation. Birge and Louveaux [2011] call such formulation a "deterministic equivalent" model since all scenarios are explicitly represented. In the following section, a decomposition strategy is proposed to explore the compound structure of the problem.

The sets and the required data are defined in Table 4.1, while the variables are summarized in Table 4.2. A set T of tasks t is distributed over stations (set S) respecting precedence relations ($Prec$). The sequencing deals with a collection of product models M. Each model m exhibits task processing times Dur_{tm}. The duration of a task of a given model can be set to zero if the product does not require such task. For the stochastic version of the problem, multiple demand scenarios $d \in D$ are considered with a probability of Pr_d. For each given demand scenario, the demand for each model (Dem_{md}) is given. Based on each scenario d, a set of product pieces $p \in \{1, ..., |MPS_d|\}$ is defined. The product set is defined with an index d because the number of pieces in a minimal part set (MPS) can differ based on the demand of each scenario. The cycle time (CT) is given and the station length is measured in relation to the equivalent distance of cycle time considering a conveyor speed of 1 and a constant length multiplier LM ($LM \cdot CT$), identical to all stations.

The first-stage decisions are based on the balancing variables x. The second stage depends on binary sequencing variables y and continuous variables U, PT, and Pos to model the utility work, processing time, and final work position for every workpiece and station for all demand scenarios.

$$\textbf{Minimize} \sum_{d \in D} \sum_{s \in S} \sum_{p \in \{1, ..., |MPS_d|\}} \frac{Pr_d \cdot U_{psd}}{|MPS_d|} \tag{4.1}$$

$$\sum_{s \in S} x_{ts} = 1 \quad \forall \, t \in T \tag{4.2}$$

$$\sum_{k \in S: k \leq s} x_{t_1 k} \geq \sum_{k \in S: k \leq s} x_{t_2 k} \quad \forall \, (t_1, t_2) \in Prec, s \in S \tag{4.3}$$

$$\sum_{t \in T} x_{ts} \cdot Dur_{tm} \leq LM \cdot CT \quad \forall \, s \in S, m \in M \tag{4.4}$$

$$\sum_{m \in M} y_{pmd} = 1 \quad \forall \, d \in D, p \in \{1, ..., |MPS_d|\} \tag{4.5}$$

$$\sum_{p \in \{1, ..., |MPS_d|\}} y_{pmd} = Dem_{md} \quad \forall \, m \in M, d \in D \tag{4.6}$$

$$PT_{psd} \geq \sum_{t \in T} x_{ts} \cdot Dur_{tm} - LM \cdot CT \cdot (1 - y_{pmd}) \quad \forall \, d \in D, m \in M, s \in S, p \in \{1, ..., |MPS_d|\} \tag{4.7}$$

$$\sum_{p \in \{1, ..., |MPS_d|\}} PT_{psd} = \sum_{t \in T} \sum_{m \in M} x_{ts} \cdot Dur_{tm} \cdot Dem_{md} \quad \forall \, d \in D, s \in S \tag{4.8}$$

$$Pos_{psd} \geq Pos_{(p-1)sd} - CT + PT_{psd} - U_{psd} \quad \forall \, d \in D, s \in S, p \in \{2, ..., |MPS_d|\} \tag{4.9}$$

$$Pos_{1sd} \geq Pos_{|MPS_d|sd} - CT + PT_{1sd} - U_{1sd} \quad \forall \, d \in D, s \in S \tag{4.10}$$

$$Pos_{psd} \geq CT \quad \forall \, d \in D, s \in S, p \in \{1, ..., |MPS_d|\} \tag{4.11}$$

$$Pos_{psd} \leq LM \cdot CT \quad \forall \, d \in D, s \in S, p \in \{1, ..., |MPS_d|\} \tag{4.12}$$

$$x_{ts}, y_{pmd} \in \{0, 1\} \tag{4.13}$$

$$U_{psd}, PT_{psd}, Pos_{psd} \in \mathbb{R}^+ \tag{4.14}$$

Table 4.1 Nomenclature of sets and data of the formulation

Sets	Meaning
T	Set of tasks t
S	Set of stations s or k ordered along the belt
M	Set of product models m
D	Set of demand scenarios d
$\{1, ..., |MPS_d|\}$	Set of pieces p of the demand scenario d
$Prec$	Set of precedence relations $(t_1, t_2), t_1, t_2 \in T$
Data	**Meaning**
Dur_{tm}	Duration time of task t of model m
Dem_{md}	Demand of model m in scenario d
Pr_d	Probability of scenario d
CT	Cycle time
LM	Length multiplier, used to define the maximal length of the stations

Table 4.2 Nomenclature of model variables. P stands for the set defined in $\{1, ..., |MPS_d|\}$

Variable	Domain	Type	Meaning
x	(T, S)	Binary	1, if task t is assigned to station s, 0, otherwise
y	(P, M, D)	Binary	1, if the pth piece is of model m for scenario d, 0, otherwise
U	(P, S, D)	$\mathbb{R}^+$	Utility work of piece p at station s and scenario d
PT	(P, S, D)	$\mathbb{R}^+$	Sum of the processing times of piece p at station s and scenario d
Pos	(P, S, D)	$\mathbb{R}^+$	Final position of the worker for piece p at station s and scenario d

The monolithic model is presented by the expressions (4.1)–(4.14). Its objective function is the minimization of the expected utility work per unit produced. Expression (4.1) is modeled as a weighted average of the utility work on all scenarios divided by the number of products in each scenario. Expressions (4.2) and (4.3) are the occurrence and precedence restrictions. The balancing part of the model is completed with ineq. (4.4). So that the products can be processed within the stations, the workload of any station cannot be longer than the station length ($LM \cdot CT$).

The sequencing part of the model assigns product models to workpieces. The assignment restrictions are modeled by equations. (4.5) and (4.6). Every workpiece must be one of the product models (equation 4.5), while the number of pieces of a given model must be equal to the demand for a given scenario (equation 4.6). The processing time variables (PT) integrate the balancing and sequencing. Expression (4.7) is a Big-M constraint assigning the processing time of workpieces to the corresponding duration of their model tasks. The sum of the processing time is also enforced by expression (4.8), which tightens the linear relaxation of the model.

Expressions (4.9)–(4.12) model the scheduling of paced assembly lines. Variable Pos controls the final position of a worker after working on a piece. This position is determined by expression (4.9) based on the position of the last piece ($Pos_{(p-1)sd}$), the time distance between the two pieces (CT), the processing time of the piece (PT_{psd}), and the employed utility work (U_{psd}). Expression (4.10) represents the cyclical sequencing link of the first and the last piece of the previous of two different replications of the MPS. Further bounds complete the formulation of the problem to assure no task is performed outside of the station borders. Expression (4.11) assures that the initial position of a piece ($Pos_{psd} - CT$) cannot be negative, as the worker can return the equivalent of CT at the line. For the maximal bound, the final work position must be within the station limit (expression 4.12). Finally, expressions (4.13) and (4.14) present the binary restrictions on variables x and y and the non-negativity of the continuous variables.

Notice that the balancing restrictions are independent of the demand scenario (no index d), while the sequencing part of the model has variables and expressions to every scenario. This structure is commonly found in stochastic problems and it is explored in the decomposition approach presented in Section 4.2.

4.2 Solution Algorithm

The proposed solution algorithm is based on a decomposition procedure that uses the structure of the problem to solve it efficiently. The procedure is based on Benders' decomposition, which is described in Section 4.2.1 and is originally proposed for linear subproblems [Benders, 1962]. The adaptation of the decomposition algorithm for subproblems with integer variables is discussed in Section 4.2.2. The proposed algorithm is then presented in Section 4.2.3.

4.2.1 Benders' Decomposition

Benders' decomposition is named after the work of Benders [1962], in which a linear or a mixed-integer linear problem is decomposed into two levels. The integer variables are kept on the first level, while linear variables can be assigned to a subproblem on the second level. The advantage of the decomposition is that the separated problems are smaller and therefore can probably be solved faster.

Stochastic optimization problems are typical applications of decomposition procedures since the multiple scenarios have a very similar structure. In the stochastic literature, Benders' decomposition is often named L-Shape method [Birge and Louveaux, 2011]. In this method, the first stage of a stochastic model is the master problem, while the subproblems are the second stage of the decision process, that is, the reaction to the uncertain events. Just as Benders' decomposition, the standard L-Shape method also requires linear variables in the subproblem.

As both, master and subproblem, are solved separately, both do not contain all information of the problem. Benders' decomposition is based on the exchange of information from both levels of the problem. For a given solution of the master problem (decisions before the outcome of the realizations), the subproblems can be solved to determine the best outcome in response to the solution. This response quantification can be given back to the master problem in form of a cutting plane. This way, the master problem with the added cut acquires information about the subproblems that are already solved. In each iteration, the master problem is solved with incomplete, but accumulating information, which is added from the solution of the subproblems. The procedure is iterated until the solutions of both problems are compatible. The reasoning behind the decomposition is that not all information of the subproblems needs to be added to the master problem to characterize the optimal solution. That is, not all feasible solutions of the master problem have to be explored by evaluating the corresponding subproblems. The expectation for good applications of Benders' decomposition is that solving significant smaller problems multiple times requires less effort than a large problem combining all variables and restrictions [Rahmaniani et al., 2017].

A general two-stage stochastic linear program can be written in the form [Laporte and Louveaux, 1993]

$$\text{Min } Z = c^T x + E_\xi \text{ Min}_y (q^T (\omega) y)$$
$$\text{s.t. } Ax = b$$
$$W y = h(\omega) - T(\omega)x \qquad (4.15)$$
$$x \in X, y \in Y,$$

in which $c \in \mathbb{R}^{n_1}$, $b \in \mathbb{R}^{m_1}$, and $A \in \mathbb{R}^{m_1 \times n_1}$, respectively, represent the objective function coefficients, the right-hand-side, and the technological matrix of the first stage problem with n_1 variables and m_1 constraints. The second-stage is defined based on a random vector ξ, which contains all stochastic components of the problem: $\xi(\omega) = (q(\omega), h(\omega), T(\omega))$. The dimension of ξ is defined accordingly by the dimensions of $q(\omega)$, $h(\omega)$, and $T(\omega)$, which are the objective function coefficient vector, the right-hand side vector, and the matrix that represents the effect of a first stage-solution (x) in the second-stage, respectively. Along with the technological matrix W, the second stage is defined based on the possible realizations ω of the random variable ξ. Matrix W can also be defined as a function of ω [Birge and Louveaux, 2011], which is not the case for the presented sequencing subproblem. Finally, E_ξ represents the expected value of the realizations of ξ. Note that this extensive formulation contains all variables and restrictions of the problem.

For the decomposition, the model can be separated into two parts. The first one retains the first-stage x variables, while the second-stage contains all scenarios. A reformulation of the master problem is presented in expressions (4.16)–(4.20). This reformulation is named *relaxed master problem* since the subproblem variables and restrictions are excluded [Rahmaniani et al., 2017]. This information, however, must be added during the algorithm's iterations. To aid in the integration, a new variable θ is added to approximate the value of $E_\xi \operatorname{Min}_y(q^T(\omega)y)$ in the objective function (eq. 4.16). The information of the subproblems can be passed on to the master problem in form of feasibility (4.18) and optimality (4.19) restrictions. The elements D_k, d_k, E_k, and e_k are the coefficients of the kth computed cut of each type in the procedure. The feasibility restrictions are useful when a solution of the relaxed master problem is shown infeasible in one of the subproblems. The coefficients D_k and d_k are chosen in a way, that the region of the master problem containing the infeasible solution is cut out. The optimality cuts (4.19) provide information from the objective function value of the subproblems. These cuts act on the variable θ, improving the objective function approximation of the master problem after each iteration.

$$\operatorname{Min} c^T x + \theta \tag{4.16}$$

$$s.t. \; Ax = b \tag{4.17}$$

$$D_k x \geq d_k \qquad\qquad k = 1, ..., s \tag{4.18}$$

$$E_k x + \theta \geq e_k \qquad\qquad k = 1, ..., t \tag{4.19}$$

$$x \in X, \theta \in \mathbb{R}. \tag{4.20}$$

The values of E_k and e_k are derived from the shadow prices of the subproblems, while D_k and d_k come from the infeasible rays of subproblem solutions [Birge and Louveaux, 2011]. Therefore, Benders' decomposition (and the L-shape method) is applied to linear subproblems in its standard form [Rahmaniani et al., 2017].

4.2.2   Combinatorial Benders' Decomposition

In the case that the subproblems contain integer variables, a different strategy for the
cuts is needed. According to a recent survey on Benders' decomposition, Rahmaniani et al., [2017] state that the cuts are still valid when based on the shadow prices
of the subproblem's linear relaxation. These cuts, however, may still leave a gap
between the integer solution and the linear relaxation. One alternative to overcome
this gap is to use the combinatorial nature of the problem to build cuts.

One of the alternatives to build cuts for integer subproblems is proposed by
Laporte and Louveaux [1993] for master problems composed of binary variables.
The cut

$$\theta \geq (\theta^k - L) \left(\sum_{i \in S_k} x_i - \sum_{i \notin S_k} x_i - |S_k| \right) + \theta^k \qquad (4.21)$$

is used to correct the value of θ based on a solution of an iteration k. S_k is the set of
the master problem variables that have value 1 in the solution of iteration k, θ^k is the
expected value of objective function after solving the subproblems, and L is a lower
bound for the objective function. Cut (4.21) assures that θ is greater than or equal
to the realized value (θ^k) when the master problem variables assume the values of
the explored solution in iteration k. This restriction is implemented using the sums
$\sum_{i \in S_k} x_i$ and $\sum_{i \notin S_k} x_i$. The restriction is only bounding when $\sum_{i \in S_k} x_i = |S_k|$ and
no variable $x_i \notin S_k$ is set to 1. If any variable assumes a different value, then θ is
only bounded to a lower bound L. If more variables assume a different value, the
cut does not restrict effectively (under the lower bound) the objective function of
the solution. Laporte and Louveaux [1993] also develop a strengthened version of
the cut using improved lower bounds based on the solution neighborhood.

A second alternative for combinatorial Benders cuts for binary master problems
is proposed by Codato and Fischetti [2006]. Their approach uses the subproblems as
a feasibility check. The approach is only applicable if the objective function of the
problem is either only based on the master problem variables or only based on the
subproblem variables. When the objective function is only based on master problem
variables, the subproblems can be used to check whether a master problem solution
is feasible for the whole model. If the solution is proven infeasible, the cut

$$\sum_{i \in C} x_i \leq |C| - 1 \qquad (4.22)$$

is added to the master problem. The set $C \subseteq X$ is a minimal infeasible subsystem
(MIS) [Codato and Fischetti, 2006]. This set contains the smallest set of variables

that causes the infeasibility. According to Codato and Fischetti [2006], the identification of the minimal infeasible subsystem may be cumbersome, so that non-minimal infeasible systems may also be used. Cut (4.22) ensures that at least one of the variables contained in C must be zero.

Cut (4.22) also works for the case in which the objective function is only based on the subproblem variables. In this case, an upper bound for minimization problems or a lower bound for maximization problems is needed. The subproblem is solved with an extra restriction, that the objective function must be strictly better than the upper bound. Hence, the master problem searches for improvements, which are verified on the subproblem level. If the solution cannot beat the bound, a cut (4.22) can be included in the master problem. This process is iterated until the master problem becomes infeasible. The latest bound cannot be improved and is optimal. Codato and Fischetti [2006] state that the use of the approach is very sensitive to the structure of the problem and may not be efficient for some applications.

There are some approaches in the assembly-line balancing literature which use the combinatorial version of the Benders' decomposition. Akpinar et al. [2017] solve the assembly-line balancing problem with task-dependent set-up times. In their problem, the tasks must be not only assigned to workstations but also scheduled within the workstation. Akpinar et al. [2017] propose a Benders' decomposition in which the master problem consists of the balancing problem and the intra-station sequencing is the subproblem. The used combinatorial cuts are based on Codato and Fischetti [2006]. If the scheduling of any station is infeasible in terms of the cycle time, the assignment of this station is cut out using cut (4.22). In this problem, the sequencing of each station can be solved independently. The MIS consists of the assignments to the infeasible station.

The second application of combinatorial Benders' decomposition in the assembly line balancing context is due to Michels et al. [2019]. In this contribution, assembly lines with multiple workers per station are optimized. Just as in Akpinar et al. [2017], multiple workers require the intra-station scheduling of operations. The approach also divides the problem into balancing as the first level and sequencing of each individual station as the second level. The proposed combinatorial cut in Michels et al. [2019] is a special version of cut (4.22). One infeasible subproblem does not necessarily mean that the assignment is infeasible, but that at least one more worker must be assigned to the station. If the station has the maximum number of available workers assigned, an infeasible subproblem shows that the assignment itself is infeasible.

Both Akpinar et al. [2017] and Michels et al. [2019] apply the combinatorial Benders' decomposition to deterministic assembly line balancing problems. In a stochastic context, Bentaha et al. [2015] survey several applications applying Ben-

ders' decomposition to the balancing of disassembly lines. These approaches, however, contain linear subproblems, so that the standard cuts based on the shadow
prices are applicable.

4.2.3 Proposed Algorithm

The decomposition strategy used for solving the problem as well as the improvements implemented in the algorithm are presented in this section.

Model decomposition

Benders' decomposition requires the problem to be split between a master problem
(MP) and one or more subproblems (SP). The master problem keeps the variables and
restrictions related to the assembly-line balancing, while the subproblem is responsible for the sequencing and scheduling of the stations. From the monolithic model
(expressions (4.1)–(4.14)), the binary variable x as well as the expressions (4.2)–
(4.4) form the master problem. A variable θ_{ds} is used as an auxiliary variable to
approximate the objective function of the subproblem for station s in scenario d. The
objective function of the master problem is then the weighted sum of approximation
variables

$$\textbf{Minimize} \sum_{d \in D} \sum_{s \in S} \frac{Pr_d \cdot \theta_{ds}}{|MPS_d|}. \tag{4.23}$$

The combinatorial cuts are defined in a set of cuts K. Set K starts empty and is
used to store the combinatorial cuts found within the algorithm iterations. These
cuts are an adaptation of inequality (4.21) [Laporte and Louveaux, 1993] and are
implemented as

$$\sum_{s \in C_k} \theta_{ds} \geq \theta_d^k \cdot \left(\sum_{(t,s) \in S_k} x_{ts} - |S_k| + 1 \right) \qquad \forall \, k \in K, d \in D. \tag{4.24}$$

On the left-hand-side of the expression, the corresponding sum of related $\theta_{d,s}$ is
used instead of a single variable θ. This is necessary for "partial combinatorial
cuts", which are based on a subset of the stations (C_k) and are described bellow.
The right-hand-side is based on the sum operator in set S_k, which contains the task-
station assignments for the stations of C_k in iteration k. C_k is the set of stations used
to generate the cut at iteration k. As it is discussed bellow, cuts can be generated
for all stations or by using a subset of stations. That is, the sum $\sum_{(t,s) \in S_k} x_{ts}$ counts

the number of variables that are equal to a given assignment S_k. The cut imposes a minimal value for the θ_{ds} variables (θ_k^d) if the assignment is exactly the same as the one explored in iteration k. For different solutions (at least one assignment is different), the cut does not constraint the θ_{ds} variables.

The subproblems are responsible for the sequencing of models within the assembly line. As each scenario d is independent of the other, $|D|$ subproblems can be defined by expressions (4.5)–(4.14). The objective is to

$$\textbf{Minimize} \quad \sum_{p \in \{1,...,|MPS_d|\}} \sum_{s \in S} U_{ps}, \tag{4.25}$$

the necessary total utility work for the sequencing of models.

Tightening the formulation: Preprocessing and valid inequalities
According to Rahmaniani et al. [2017], one approach to deal with integer subproblems in Benders' decomposition is to use all available information of the linear relaxation of the subproblem and use combinatorial cuts to narrow the gap between the relaxation and the integer solution. Therefore, it is important to tighten the formulation so that the relaxation is as close as possible to the integer model [Fischetti et al., 2017].

In the next paragraphs, the development of preprocessing, valid inequalities, and alternative combinatorial cuts are presented. The descriptions of the *Balancing preprocessing, Valid inequalities based on the linear relaxation, Unavoidable idle time*, and *Partial combinatorial cuts* are taken from Sikora [2021]. A further improvement is to use local search in the master problem to find better solutions quickly. The integrated local search procedure is not contained in Sikora [2021] and is therefore only proposed here.

Balancing preprocessing
The preprocessing consists of identifying beforehand assignment variables that cannot assume the value 1 in the optimal solution. For the simple assembly-line balancing problem, Patterson and Albracht [1975] define the earliest (E_t) and latest (L_t) station a task t can be assigned to. The preprocessing is based on equations (4.26) and (4.27) and uses the sum of the processing times of the task's predecessors and successors to define its bounds. In equation (4.26), $Prec^+$ is the set of all direct and indirect precedence relations, so that $\sum_{(i,t) \in Prec^+} Dur_i$ is the sum of the processing time of all predecessors of task t and $\lceil \cdot \rceil$ is the ceiling function. The numerator is the minimal amount of time that is needed to process a task and its predecessors, which divided by the cycle time returns the smallest number of

stations required for its assignment. Similarly, equation (4.27) is used for the upper bound on the line position.

$$E_t = \left\lceil \frac{Dur_t + \sum_{(i,t) \in Prec^+} Dur_i}{CT} \right\rceil \qquad \forall t \in T \qquad (4.26)$$

$$L_t = |S| - 1 + \left\lceil \frac{Dur_t + \sum_{(t,i) \in Prec^+} Dur_i}{CT} \right\rceil \qquad \forall t \in T \qquad (4.27)$$

Patterson and Albracht's 1975 preprocessing is based on an assembly line with only one model. The adaptation of these bounds for the proposed problem has to consider the multiple models and the demand scenarios. A first preprocessing possibility is to consider an average model. E_t^{avg} and L_t^{avg} are defined as the earliest and latest station based on the tasks' average durations using the same expressions of equations (4.26) and (4.27). This average duration is the weighted average based on the demand for the products and the scenarios' probabilities. This bound is only valid under the assumption that the line can operate for the average model (the single model equivalent, for which the duration of each task is equal to the expected duration of the task) without requiring utility work. This is not unrealistic, since the assembly line would be undersized if it could not handle the planned workload on average.

Treating models individually also provides bounds for the assignment variables. In inequality (4.4) of the monolithic model, it is assumed that all models can be produced within the station length. That is, no processing time equivalent may be larger than the size of the station ($LM \cdot CT$), otherwise, the production would be infeasible even with utility work. A model-specific Earliest Station for model m

$$E_t^m = \left\lceil \frac{Dur_{tm} + \sum_{(i,t) \in Prec^+} Dur_{im}}{CT \cdot LM} \right\rceil \qquad \forall t \in T, m \in M$$

and Latest Station

$$L_t^m = |S| + 1 - \left\lceil \frac{Dur_{tm} + \sum_{(t,i) \in Prec^+} Dur_{im}}{CT \cdot LM} \right\rceil \qquad \forall t \in T, m \in M$$

are valid for the problem if the station length is considered as the loading limit of each station.

As any assignment (t, s) outside the interval $[E_t^m, L_t^m]$ for any model m is infeasible, the intersection of the average and the model-specific bounds can be computed. The resulting sets are used to specify the balancing variables, which already excludes many variables that could not result in a feasible answer from being set to a value of 1.

Valid inequalities based on the linear relaxation

As already mentioned, Rahmaniani et al. [2017] describe in their survey the use of the shadow prices of the linear relaxation of the subproblem in Benders' decomposition. For the presented problem, however, it is not necessary to solve the linear relaxation of the subproblem, since the deriving cut can be simply defined as

$$\theta_{ds} \geq \sum_{t \in T} \sum_{m \in M} Dur_{tm} \cdot Dem_m \cdot x_{ts} - |MPS_d| \cdot CT \quad \forall d \in D, s \in S. \quad (4.28)$$

The development and proof of the expression are given in the online appendix A of Sikora [2021].

One interpretation for cut (4.28) is that utility work is necessary if the assignment of any station is overloaded in any demand-scenario. That is, the sum of the processing times is larger than the available processing time ($|MPS_d| \cdot CT$), resulting in utility work independent of the sequence. This cut is intuitive and can be added directly to the master problem.

Unavoidable idle time

The cut based on the linear relaxation of the subproblem can be improved if idle time can be identified in the master problem. As each new product flows into the station after a cycle time (CT time units), very short processing times that result in idle time may be detected in the master problem already. For this cut, the non-negative variable I_{ms} is used as an auxiliary to measure the unavoidable idle time of a model m in station s. A lower bound on the idle time can be calculated by

$$I_{ms} \geq (2 - LM) \cdot CT - \sum_{t \in T} Dur_{tm} \cdot x_{ts} \quad \forall m \in M, s \in S. \quad (4.29)$$

For the explanation of the cut, an example is illustrated in Figure 4.2. After finishing a model, the worker returns the equivalent of CT time units to the beginning of the station. Hence, the maximal start position is $LM \cdot CT - CT$. If the processing time of any model is shorter than the difference between this point and the cycle time mark ($\sum_{t \in T} Dur_{tm} \cdot x_{ts} \leq CT - (LM \cdot CT - CT) = (2 - LM) \cdot CT$), idle time occurs independently of the sequence. Therefore, cut (4.29) implements a lower bound on the idle time and can be included in the master problem formulation. Note that this cut is only useful if the line length is smaller than twice the cycle time

($LM < 2$). Otherwise, at least two products would be at the station at any given time and the lower bound on the idle time given by the cut is zero.

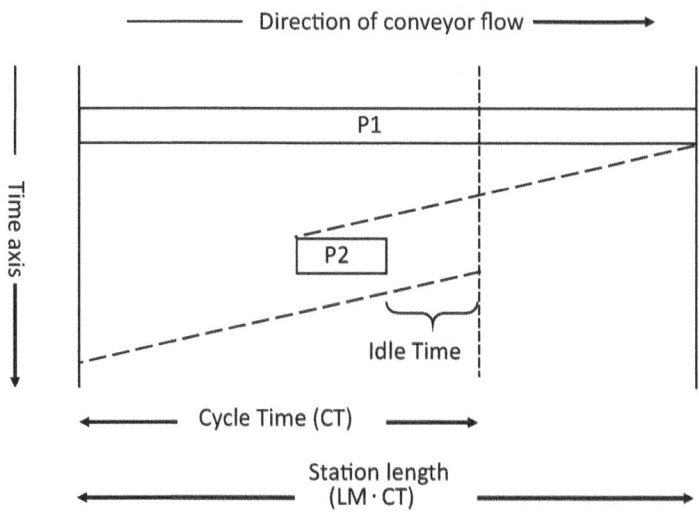

Figure 4.2 Example of assignment that invariably causes idle time. Figure from Sikora [2021]

The unavoidable idle times due to the lower bound restriction of cut (4.29) can be integrated into the cut based on the linear relaxation. Expression

$$\theta_{ds} \geq \sum_{t \in T} \sum_{m \in M} Dur_{tm} \cdot Dem_m \cdot x_{ts} + \sum_{m \in M} Dem_m \cdot I_{ms} - |P| \cdot CT \quad \forall d \in D, s \in S$$

(4.30)

uses the information of the idle times to strengthen the cut, since the idle times reduce the available time for processing.

Partial combinatorial cuts

Although the preprocessing and the linear-relaxation cut help bridging the gap between the objective function approximation on the master problem and the realization of the subproblems, they are not enough to define the formulation of the problem. As cited in Laporte and Louveaux [1993] as well as Codato and Fischetti

[2006], combinatorial cuts may be needed to represent the whole objective function landscape. These combinatorial cuts are, however, very local, in a way that they only apply to the explored solution or solution subset. Cut (4.24) built based on the assignment of all stations would only correct the approximation value of the assignment at iteration k. For all other possible assignments in which at least one variable differs, the restriction is not binding.

In order to generate cuts that affect the approximation of other assignments, cuts based on partial assignments are proposed. For these cuts, sequencing subproblems are solved for a subset of the workstations, which form the workstations set C_k defined in expression 4.24. The solutions to these problems are not representative of the whole sequencing problem, since the sequence has to be the same for all stations. However, considering a subset of stations provides a lower bound for the objective function of the whole line. Therefore, the lower bound found for the partial assignments can also be given as a cut to the master problem. Since cut (4.24) is already written in terms of a subset of stations, the same cut can be used as the "complete" (based on all stations) as well as the partial (based on a subset of stations) cut.

The proposed cut is illustrated with a small numeric instance given in Table 4.3. The same example is used in the description of Sikora [2021]. For the instance, four stations and four models are given (M1–M4). For simplicity, only one demand scenario with an MPS of 1-1-1-1 is considered. Further parameters are a cycle time of five time-units and a station length of seven time-equivalent units. Table 4.3 contains the processing time of the eight tasks and three balancing solutions So1, So2, and So3.

The first balancing solution (So1) corresponds to the station-wise processing times presented in Table 4.4. This solution is feasible for the master problem and has an approximated objective function value of zero. That is, no utility work is expected. By solving the sequencing problem, the optimal sequencing solution $1 - 3 - 2 - 4$ is obtained. This solution, however, contains one unit of utility work. The combinatorial cut (4.24) becomes in this case

$$\theta_1 + \theta_2 + \theta_3 + \theta_4 \geq 1 \cdot (x_{11} + x_{21} + x_{32} + x_{42} + x_{53} + x_{63} + x_{74} + x_{84} - 8 + 1). \quad (4.31)$$

This cut can be incorporated into the master problem to correct the approximation of the objective function value for this assignment. A possibly optimal solution of the master problem with the cut (4.31) is represented by So2 in Table 4.3. This balancing solution also has an objective function approximation value of zero, which is not corrected by cut (4.31). The subproblem sequencing optimal solution is also $1 - 3 - 2 - 4$ containing one unit of utility work. In this example, cut (4.31) could

Table 4.3 Task durations and solutions for an example instance. The columns 'Solution' contains the station to which each task is assigned

Task	Processing Time				Solutions		
	M1	M2	M3	M4	So1	So2	So3
1	4	2	2	4	1	1	1
2	2	1	1	3	1	1	2
3	3	2	1	1	2	2	1
4	4	4	3	2	2	2	2
5	1	1	2	3	3	3	3
6	2	3	4	4	3	4	3
7	5	4	2	2	4	3	4
8	2	3	1	1	4	4	4

Table 4.4 Processing times per station and model for the first balancing solution

Model	Station			
	1	2	3	4
1	6	7	3	7
2	3	6	4	7
3	3	4	6	3
4	7	3	7	3

not be extended to other assignments, as the information (one unit of utility work is caused) is only valid for exactly that assignment.

The partial cuts present an alternative and use lower bounds to generate cuts. Solving the sequencing of each station of So1 individually results in the sequence $1 - 2 - 4 - 3$ for the first station and $1 - 3 - 2 - 4$ for the others containing no utility work. Combining the first and the second station in a partial subproblem results in the optimal sequence $1 - 3 - 2 - 4$ with one unit of utility work. This solution is a lower bound for the sequencing problem, since considering more stations would result in the same one unit of utility work or more. The resulting cut

$$\theta_1 + \theta_2 \geq 1 \cdot (x_{11} + x_{21} + x_{32} + x_{42} - 4 + 1) \qquad (4.32)$$

provides a valuable piece of information to the master problem: the unit of utility work from So1 is caused by the bad correspondence of the assignments of stations 1 and 2. Cut (4.32) contains fewer variables, which potentially form different assignment solutions in the master problem. Solution So2 of Table 4.3, for instance, also contains the same assignments. This solution would not be explored by the master problem in the presence of cut (4.32), reducing the number of master problem solutions that must be explored in the algorithm. The optimal solution for the problem is So3, which does not need utility work for the sequence $1 - 3 - 2 - 4$.

The partial combinatorial cuts are very useful if utility work on subsets of stations can be identified. The partial combinatorial cuts are also valid for other assignments so that the information helps the quality of the master-problem approximation of the objective function. A related approach is described in Fischetti et al. [2016] in the context of relaxing fixed-cost restrictions. The relaxed problems are also lower bounds of the problem, for which cuts can be defined.

The strategy of solving subsets of the problem may be applied to other optimization problems as well. If the optimal solution of the subset of the problem produces a lower bound (for minimization problems), the approach may be adequate. In the assembly-line context, decomposing based on stations results in a lower bound. Solving the sequencing problem for a subset of models, however, provides an upper bound of the problem, when the subsequences are then merged together. Although this may be a valid method to obtain valid sequencing solutions, the objective function value is an upper bound and may not be used for cut generation.

It is important to notice that the partial subproblems are as complex as the complete subproblem, so that it may not always be advantageous to solve multiple partial subproblems. For the proposed problem, the sequencing problems with a small subset of stations are solved quickly with commercial solvers and have shown themselves a viable approach to generate cuts.

Integrated local search
An iterative decomposition procedure consists of solving master and subproblems repeatedly. Rahmaniani et al. [2017] advise a good balance between the solution time that is required by the master problem and the subproblems. It is, however, common, that the master problem requires longer solution times to produce feasible solutions. One of the approaches to reduce the solution time of the master problem is to use heuristics to obtain solutions [Costa et al., 2012; Rahmaniani et al., 2017; Caserta and Voβ, 2020].

In this section, an idea presented by Caserta and Voβ [2020] is used to accelerate the convergence of the algorithm. Instead of solving the whole master problem to find the solutions, an integrated local search is used on the master problem to obtain feasible solutions fast.

The local search can be performed in the mathematical model that defines the master problem adding restrictions of the form

$$\sum_{(t,s)\in S_w} x_{ts} \leq |T| - 1 \qquad (4.33)$$

$$\sum_{(t,s)\in S_w} x_{ts} \geq |T| - R. \qquad (4.34)$$

In the expressions, S_w represents the task-station pairs of the balancing solution of iteration w, and R is a constant. Constraints (4.33) and (4.34) restrict the balancing variables to a solution that are similar to but different from solution S_w. The parameter R is the search radius and is equivalent to the number of assignments that can be changed in the local search. With these restrictions, the search space is strongly restricted to the region near to a known solution. This way, a quick search of the neighborhood may provide fast feasible solutions and accelerate the overall convergence of the decomposition [Caserta and Voβ, 2020].

Description of the algorithm
In the original Benders' decomposition by Benders [1962], both the master and subproblems are solved to optimality in every iteration of the algorithm. The optimality of the master problem is, however, not required for the solution of the subproblems and the generation of cuts. According to the survey of Rahmaniani et al. [2017], the time required to solve the master problem can take more than 80% of the solution time, so that the complete solution of the master problem can be very costly. Fischetti et al. [2017] present a "modern implementation" of Benders' decomposition, in which the master problem is solved only once. Several universal solvers support the use of callbacks, that add cuts dynamically to the formulation. This way, Benders' decomposition cuts can be added "on the fly" for every feasible solution found in the master problem, increasing the speed of the algorithm [Codato and Fischetti, 2006; Costa et al., 2012; Akpinar et al., 2017; Fischetti et al., 2017; Michels et al., 2019].

The algorithm is described in Algorithm 4.1. The master problem is solved with a single call in a solver and the other functions are called every time a new incumbent solution is found. That is, a feasible balancing solution for the master problem with a corresponding approximation for the utility work. In the initialization, the global upper bound UB and the master problem's lower bound LB are set. The search of the master problem continues until the global upper bound and the master problem's lower bound are the same, providing the optimal solution.

For each solution found in the master problem, the corresponding subproblems are solved in the routine 'ExploreSolution'. In this process, combinatorial cuts are added to the master problem, which are considered for the next solutions. Before returning to the master problem, a given number of iterations of a local search is performed for each solution found for the master problem.

Algorithm 4.1: Combinatorial Benders's decomposition

Result: Optimal solution
$LB = 0$; $UB = \infty$;
while $UB > LB$ **do**
    Sol = FeasibleSolution(MasterProblem);
    LB = MasterProblem.LowerBound;
    $i = 0$;
    while $i < No.$ *Iterations Local Search* **do**
        ExploreSolution(Sol, UB, $Depth$);
        Sol = LocalSearch(Sol, R);
        $i + +$;
    end
end

The exploration of the solution in the subproblem level is described in Algorithm 4.2. The subproblems are explored in two levels. The first optional part of the algorithm depends on a parameter $Depth$, which controls how the stations are combined in the partial subproblems. A depth of one represents solving each station individually, a depth of two combines two stations, etc. The number of combinations is also part of the algorithm design. Options are to solve all combinations for a given number of stations or just a subset of them. The solution of the partial problems results in a lower bound for each depth i (LB_i^l). If the lower bound of the given solution surpasses the global upper bound (UB), the solution can be cutoff with combinatorial cuts and does not need to be explored further. If the solution is not cut off based on the partial subproblems, the subproblems are explored in the second level of the algorithm (after the while function). The complete subproblems are solved (SolveSubProblems) resulting in a local upper bound (UB^l). If the local upper bound is better than the previous best solution (UB), the global upper bound is updated. Every time any utility work is identified, cuts are added to the master problem.

Algorithm 4.2: ExploreSolution(Sol, UB, $Depth$)

Result: Solve the subproblems and add cuts

$i = 1$; **while** $i \leq Depth$ **do**

$\quad\quad LB_i^l$ = SolvePartialSubProblems(Sol, $Depth$);

$\quad\quad$ AddCuts;

$\quad\quad$ **if** $LB_i^l >= UB$ **then**

$\quad\quad\quad$ | exit routine

$\quad\quad$ **else**

$\quad\quad\quad$ | i++

$\quad\quad$ **end**

end

UB^l = SolveSubProblems(Sol);

AddCuts;

if $UB > UB^l$ **then**

$\quad$ | $UB = UB^l$

end

For the results described in the next section, some values for the implementation parameters are set. The $Depth$ is set to 3 so that the combinations of up to three workstations are solved as partial subproblems. Not all combinations are explored, but the selection is made as a minimal cover of all stations. For simplicity, the stations are gathered in sequential order (that is, 1–2, 3–4, ...). The search radius (R) of the local search is set to 6 tasks, that is, at most 6 tasks can be reassigned in each iteration of the local search. The number of iterations of the local search is set to 3. Finally, a time limit of 60 seconds is used for each iteration of the local search, so that the solution procedure does not get stuck by chance in problems that may be too time-consuming to solve.

4.3   Tests and Results

The algorithm's results and performance are discussed in this section. As the algorithm is already explored in Sikora [2021], the results of this manuscript are meant to be complementary to those of the paper. For the datasets described in Section 4.3.1, the effects of the partial cuts (Section 4.3.2) and the local search (Section 4.3.3) are analyzed.

Table 4.5 Parameters used in the generation of the dataset

Parameter	Values
Number of tasks ($|T|$)	50
Number of models ($|M|$)	10
Number of stations ($|S|$)	10
Number of demand scenarios ($|D|$)	5
Ordering strength (OS)	20% / 60% / 90%
Average processing load (PL)	90% / 95%
Length multiplier (LM)	120% / 150% / 200%

4.3.1   Dataset

The instances used for the tests are based on two datasets. The first is contained in
Sikora [2021]. This dataset has 80 instances and is built based on the single model
instances from Otto et al. [2013]. For each instance, the number of used product
models ($|M|$) is 10, so that ten single model instances are used to generate one
mixed-model instance. The parameters for the instance generation are described
in Table 4.5. All instances are based on the medium-sized instances of Otto et al.
[2013], containing 50 tasks ($|T|$). The number of stations ($|S|$) is fixed to 10, while 5
demand scenarios are used ($|D|$). The variable parameters are the ordering strength
(OS), the average processing load (PL), and the station length multiplier (LM). The
OS is a measure of how many precedence relations exist from all possible relations
[Otto et al., 2013]. The parameter PL is the expected load of the line. A value
of 90% means that on average, workers need to operate 90% of the time to finish
the production considering every model of every demand scenario. A production
level of 100% is virtually impossible, since there may be idle times generated by
the balancing or the sequencing of the different products. Finally, the length of the
station is defined by the length multiplier ($LM \geq 1$). As above, the station length
is $LM \cdot CT$, which may be the equivalent of 120%, 150%, or 200% of the cycle
time. These parameters can be interpreted as the problem flexibility. The ordering
strength is related to the task assignment flexibility; the average processing load is
an efficiency flexibility (or how tight the line sizing is); the length multiplier affects
the station length and therefore the sequencing flexibility.

The parameter PL is directly related to the cycle time of each instance. The
formula for determining the cycle time is given by

Table 4.6 Model demand for each of the scenarios in every instance of the dataset

Scenario					Model No.						Prob.
	1	2	3	4	5	6	7	8	9	10	
1	1	1	1	1	1	1	1	1	1	1	0.2
2	3	3	2	2	0	0	0	0	0	0	0.2
3	0	0	0	0	4	3	3	0	0	0	0.2
4	2	2	2	1	0	0	0	1	1	1	0.2
5	0	0	0	0	2	2	2	2	1	1	0.2

$$CT = \left\lceil \frac{\sum_{t \in T} \sum_{m \in M} \sum_{d \in D} Dur_{tm} \cdot Dem_{md} \cdot Pr_d}{PL \cdot NS} \right\rceil .$$

The five demand scenarios are described in Table 4.6. For this dataset, the probability of each demand scenario is considered to be equal (20%). The demand for each model varies in each scenario. The scenario generation is inspired by the work of Chica et al. [2013], who define production plans containing the demand of the individual products. Scenario 1 has the same demand for each model. Scenarios 2 and 3 have a demand for only products of family 1 (models 1, 2, 3, and 4) or family 2 (models 5, 6, 7), respectively. Scenarios 4 and 5 present a mix of families 1 and 3 (models 8, 9, and 10) and 2 and 3, respectively.

The second dataset contains instances similarly generated, but uses slightly different settings. This dataset is used for a full enumeration in Subsection 4.3.2 and therefore is built with only 8 stations. Furthermore, PL values of 80 % and 100 % are also tested.

The instances and results of the first dataset are available in the online supplementary material of Sikora [2021]. Furthermore, supporting files of this chapter can be found at https://www.bwl.uni-hamburg.de/or/team/celso-sikora.html or https://celso-sikora.com/publication-list.

4.3.2 Strength of the Partial Cuts

This section brings the results of tests on the effect of the partial cuts. The instances used for the test are contained in the second dataset, containing instances with 8 stations.

The experiment to measure the strength of the partial cuts is developed by building all possible cuts of a given depth and comparing the lower bound with the upper bound of the assignment solution. Table 4.7 contains the results for the solutions obtained for 33 instances. The 33 instances were solved with the algorithm without the partial cuts and local search enhancement and all intermediary feasible assignments were stored. For each of the 7,880 balancing solutions (of the 33 instances), the subproblems consisting of all combinations of stations are solved. The values of the ratio of the lower bound and the upper bound are reported in Table 4.7. This value ranges from 0 to 1, in which 1 means that the lower and upper bound are identical.

The solutions are displayed based on the instance that is used to generate them. The 33 lines of Table 4.7 represent all 7,880 solutions found during the computation of the instances. Column "No. Sol." displays how many feasible solutions are found for instance. The Columns LB_0 to LB_5 represent the strength of the partial cuts for different levels of the $Depth$ parameter. LB_0 is the lower bound based just on the linear relaxation cut. LB_1 is the relative bound obtained after sequencing all stations individually. LB_2 (and the other ones) shows the relative bound when solving the sequencing problem of every possible combination of two (or more) stations. The complete enumeration of the cuts is performed up to the combination of five stations. Note that these bounds require all combinations of stations, which may not applicable or not beneficial in a search algorithm due to excessive computation time.

The results demonstrate the effects of the parameter level used to build the instance. Long stations with $LM = 2$, for instance, provide large sequencing flexibility. For this case, the approximation of the master problem and the result of the subproblems are identical for the tested instances. Therefore, the sequencing of these instances does not have large importance. Implementing such systems, however, may be costly because they require a lot of space.

The study helps to identify the instances in which the master problem can badly approximate the objective function. From Table 4.7, the instances with short station lengths ($LM = 1.2$) present a large difference between lower and upper bound (in particular for $PL = 0.9$). Furthermore, it can be seen that the improvement of the gap decreases with the addition of further stations. From this analysis and experimental tests, the combinations with up to 3 workstations are selected for the design of the solution algorithm, as implemented in Sikora [2021].

Table 4.7 Strength of the partial cuts for sample solutions

OS	PL	LM	No. Sol.	LB_0	LB_1	LB_2	LB_3	LB_4	LB_5
0.2	0.8	1.2	46	0.552	0.876	0.924	0.970	0.995	1
0.2	0.8	1.5	11	0.954	0.955	0.983	1	1	1
0.2	0.8	2.0	3	1	1	1	1	1	1
0.2	0.9	1.2	1434	0.433	0.737	0.826	0.927	0.970	0.990
0.2	0.9	1.5	154	0.727	0.902	0.979	0.999	1	1
0.2	0.9	2.0	14	1	1	1	1	1	1
0.2	0.95	1.2	470	0.528	0.879	0.909	0.946	0.969	0.982
0.2	0.95	1.5	168	0.857	0.945	0.973	0.988	0.997	1
0.2	0.95	2.0	33	1	1	1	1	1	1
0.2	1.0	1.5	629	0.858	0.923	0.966	0.987	0.996	0.998
0.2	1.0	2.0	36	1	1	1	1	1	1
0.6	0.8	1.2	20	0.798	0.958	0.982	0.991	0.997	1
0.6	0.8	1.5	10	0.940	0.976	1	1	1	1
0.6	0.8	2.0	6	1	1	1	1	1	1
0.6	0.9	1.2	1359	0.370	0.728	0.865	0.942	0.974	0.990
0.6	0.9	1.5	15	0.999	0.999	1	1	1	1
0.6	0.9	2.0	15	1	1	1	1	1	1
0.6	0.95	1.2	67	0.632	0.934	0.955	0.978	0.987	0.994
0.6	0.95	1.5	369	0.767	0.955	0.973	0.988	0.997	0.999
0.6	0.95	2.0	32	1	1	1	1	1	1
0.6	1.0	1.5	783	0.965	0.994	0.997	0.997	0.999	1
0.6	1.0	2.0	20	1	1	1	1	1	1
0.9	0.8	1.2	123	0.677	0.945	0.980	0.993	0.999	1
0.9	0.8	1.5	19	0.994	1	1	1	1	1
0.9	0.8	2.0	5	1	1	1	1	1	1
0.9	0.9	1.2	932	0.347	0.728	0.834	0.915	0.966	0.993
0.9	0.9	1.5	30	0.910	0.964	0.992	0.999	1	1
0.9	0.9	2.0	6	1	1	1	1	1	1
0.9	0.95	1.2	753	0.755	0.944	0.96	0.977	0.989	0.994
0.9	0.95	1.5	114	0.827	0.941	0.971	0.998	0.999	1
0.9	0.95	2.0	23	1	1	1	1	1	1
0.9	1.0	1.5	154	0.981	0.990	0.995	0.997	0.999	1
0.9	1.0	2.0	27	1	1	1	1	1	1

4.3.3 Effect of The Local Search

The 80 instances of the first dataset are solved with a time limit of 3,600 seconds for this test. Benders' decomposition algorithm is implemented in Visual Basic 15.0 and the master and subproblems are solved using the commercial solver Gurobi 8.1. The tests are performed on an Intel i7 8700K processor with 6 cores at 4.0 GHz and 32 GB of RAM.

The master problem is solved only once. For each feasible solution found, callbacks are used to generate subproblems, solve them, and add combinatorial cuts. The cuts are then added as 'lazy cuts' as suggested by Fischetti et al. [2017].

The results of this section extend the ones of Sikora [2021], in which the local search is integrated into the algorithm. In Table 4.8 the results are summarized. Each row of the table contains information from one version of the solution method. As a comparison, the monolithic model (the MILP model without any decomposition) is also displayed. In the summary of results, the average upper bound (UB), lower bound (LB), number of instances solved to optimality (Opt) are displayed. The solution time is divided into the solution time used for solving the master problem (Master) and the subproblems (Sub). As the monolithic model does not have the division of master and subproblems, the average of the total solution time is displayed. The columns named under # Nodes contain the average number of incumbent and cutoff nodes found by the algorithm. 'L1' refers to nodes that are proven not to be optimal after applying combinatorial cuts based on one station only. Similarly, the 'L2&3' column has the average number of nodes cut off after applying combinatorial cuts combining 2 or 3 stations. The average number of nodes cut out after solving the complete subproblems (all stations) is displayed under the column 'Full'. The incumbent nodes (Inc) were not cut out by the combinatorial cuts and presented a better solution than the known upper bound during the search. Finally, the number of generated combinatorial cuts is displayed under column '# Cuts'. The cuts are divided into 'L1', 'L2&3', and 'Full'-cuts, representing the combinatorial cuts using 1 station, 2 and 3, and all stations, respectively.

From the results of Table 4.8, it can be noticed that the integration of the local search improves the algorithm. The difference is, however, not large, since only one extra instance can be solved with the addition of the local search. The average value of the upper bound is reduced from 201.98 to 193.00, which shows that the algorithm of Sikora [2021] can be further improved with the local search addition. The lower bound with the local search is worse than the one of the base algorithm. This difference is due to only one instance (instance number 49), for which the lower bound was still evaluated at 0 after 3,600 seconds of solution time for the algorithm with local search. No reason besides a different behavior of the solver can be identified.

Table 4.8 Summary of the algorithm results

Method	UB*	LB	Opt	Sol. time Master	Sub	# Nodes L1	L2&3	Full	Inc	# Cuts L1	L2&3	Full
Monolithic	254.61	145.45	49	1499.2								
Base Algorithm	201.98	168.08	66	727.1	45.1	0.7	1.4	6.0	15.0	12.7	15.7	22.7
With LS – Radius 3	197.42	161.82	67	816.9	28.5	0.4	0.9	5.1	7.9	8.3	12.0	16.3
With LS – Radius 10	193.00	161.26	67	978.7	33.0	0.2	0.9	5.4	8.4	7.5	14.5	18.8

*The upper bound column (UB) contains the average of only 77 of the 80 instances, since not all methods (Base Algorithm and Monolithic Model) could find a feasible solution for all instances.

The average results of each instance group are described in Tables 4.9, 4.10, and 4.11. As already discussed in the results of Sikora [2021], most of the time used by the search algorithm is spent on the master problem. According to Rahmaniani et al. [2017], algorithm designers should aim at a good balance between the time spent in the master and the subproblems. The idea of integrating the local search into Benders' algorithm [Caserta and Voβ, 2020] is an attempt to shift the focus to the subproblem since feasible master-problem solutions would quickly be obtained by local search. This expected behavior is, however, not observable in the test results. The solved instances show an even more uneven distribution of the effort to solve master and subproblems. In a close inspection of the solution files, it is observable that the algorithm with local search can find good quality solutions earlier than the base algorithm. These good quality solutions prune several assignments that would be explored by the base algorithm but do not need to be tested based on a better upper bound value. Therefore, this at first glance counterintuitive effect of reducing the number of explored nodes occurred by integrating local search into the algorithm.

The results displayed in Tables 4.10 and 4.11 are related to the algorithm with 3 and 10 iterations of the local search algorithm, respectively. The difference in the number of iterations has direct implications in the average solution time required for the algorithm. For 14 of the 16 instance groups (sets of instances with same parameter constellation), the version of the algorithm with 3 iterations requires less time. The quality of the solution based on their upper bound varies. For some instances, the method with 3 iterations performs better, while the opposite occurs using 10 iterations for other instances. In general, the local search improves the solution quality of the algorithm. Increasing the number of iterations of the local search, however, does not necessarily lead to better solutions.

Table 4.9 Results of the base algorithm without local search as implemented in Sikora [2021]. Each line contains the average results of 5 instances of each combination of parameters

Parameters			UB	LB	Opt	Sol. time		# Nodes				# Cuts		
OS	PL	LM				Master	Sub	L1	L2&3	Full	Inc	L1	L2&3	Full
0.2	0.9	1.2	0	0	5	60.5	51.9	1.4	4.2	8.4	13.6	23.2	25.4	35.8
0.2	0.9	1.5	0	0	5	1.2	14.6	0.2	0.2	0.8	12.4	3.6	3.8	10
0.2	0.9	2	0	0	5	1.2	1.9	0	0	0	11.6	0	0	0
0.2	0.95	1.2	338.78	0	0	3441.2	158.9	1.8	4.3	15.8	11.5	68.0	76.8	71.0
0.2	0.95	1.5	0	0	5	560.6	38.7	3.4	5.2	9.2	22.8	16.6	20.2	28.8
0.2	0.95	2	0	0	5	17.1	3.3	0	0	0	17.4	0	0	0
0.6	0.9	1.2	11.52	0	3	1627.9	89.3	0.8	1.8	16.4	12.4	30.0	25.8	57.4
0.6	0.9	1.5	0	0	5	3.1	11.3	0.4	0.6	2.6	14.6	4.4	6.4	9.8
0.6	0.9	2	0	0	5	1.9	2.4	0	0	0	14.2	0	0	0
0.6	0.95	1.2	810.9	346.55	2	2853.4	223.3	0.6	1.4	16.8	8.4	56.6	68.2	77.2
0.6	0.95	1.5	70.12	26.02	2	2091.2	106.6	2	5.4	26.8	23.4	10.2	36.2	69.2
0.6	0.95	2	30.72	23.68	4	923.8	4.3	0	0	0	20	0	0	0
0.9	0.9	1.5	90.02	90.02	5	2.6	12.2	0	0.4	1.4	13	2.0	1.8	4.6
0.9	0.9	2	62.32	62.32	5	4.3	3.1	0	0	0	16.4	0	0	0
0.9	0.95	1.5	1133.9	1133.9	5	2.9	28	0.2	0	1.4	11	1.4	1.6	14.0
0.9	0.95	2	1006.7	1006.7	5	8.7	3.8	0	0	0	19.6	0	0	0
Avg. / Total			220.71*	168.08	66	725.1	47.1	0.67	1.47	6.22	15.14	13.5	16.63	23.61

*The average of the upper bounds refers to 79 instances with a feasible solution. No feasible solution was found in 3,600 s for one instance.

Table 4.10 Results of the algorithm with local search with 3 iterations per feasible solution

Parameters						Sol. time		# Nodes				# Cuts		
OS	PL	LM	UB	LB	Opt	Master	Sub	L1	L2&3	Full	Inc	L1	L2&3	Full
0.2	0.9	1.2	0	0	5	152.0	13.4	1.4	1.6	8.0	4.8	4.6	4.6	14.4
0.2	0.9	1.5	0	0	5	23.5	2.9	0	0	0	4	0.6	0.4	1.4
0.2	0.9	2	0	0	5	8.5	0.9	0	0	0	5.6	0	0	0
0.2	0.95	1.2	241.14	0	0	3543.0	90.7	0.5	3.0	6.3	7.8	35.8	42.8	41.8
0.2	0.95	1.5	0	0	5	579.7	12.8	2.2	2.6	3.8	7.8	7.4	8.2	15.8
0.2	0.95	2	0	0	5	557.5	1.7	0	0	0	8.4	0	0	0
0.6	0.9	1.2	8.10	0	4	1396.2	67.8	0.6	3.8	30.2	6.8	9.2	27.6	54.8
0.6	0.9	1.5	0	0	5	23.7	2.8	0.4	0	1.0	4.2	2.0	0.8	2.6
0.6	0.9	2	0	0	5	15.5	1.2	0	0	0	6.4	0	0	0
0.6	0.95	1.2	778.90	234.27	2	2628.2	201.6	0.2	1.4	15.8	8.6	62.6	78.8	78
0.6	0.95	1.5	62.44	33.90	2	2324.3	45.2	0.8	3.0	16.0	16.0	14.0	31.4	44.6
0.6	0.95	2	29.64	27.92	4	1726.2	3.4	0	0	0	14.8	0	0	0
0.9	0.9	1.5	90.02	90.02	5	17.9	5.6	0	0	0.4	6.8	1.6	2.8	4.2
0.9	0.9	2	62.32	62.32	5	20.3	1.4	0	0	0	8.2	0	0	0
0.9	0.95	1.5	1133.94	1133.94	5	19.6	14.8	0	0	1.6	5.2	1.0	2.2	9.6
0.9	0.95	2	1006.68	1006.68	5	27.9	2.3	0	0	0	11.4	0	0	0
Avg. / Total			212.97*	161.82	67	816.9	28.5	0.4	0.9	5.1	7.9	8.3	12.0	16.3

*The average of the upper bound refers to 79 instances to enable the comparison between algorithms.

Table 4.11 Results of the algorithm with local search with 10 iterations per feasible solution

Parameters			UB	LB	Opt	Sol. time		# Nodes				# Cuts		
OS	PL	LM				Master	Sub	L1	L2&3	Full	Inc	L1	L2&3	Full
0.2	0.9	1.2	0	0	5	299.2	22.4	1.6	3.6	14.6	6.4	6.2	7.4	23.4
0.2	0.9	1.5	0	0	5	32.8	2.9	0	0	0	4.2	0.6	0.4	1.6
0.2	0.9	2	0	0	5	14.8	0.8	0	0	0	5.0	0	0	0
0.2	0.95	1.2	154.27	0	0	3639.4	179.1	1.0	3.0	20.8	11.3	33.5	110.5	88.3
0.2	0.95	1.5	0	0	5	1422.1	9.0	0.2	0.6	0.8	7.2	3.4	5.2	12.0
0.2	0.95	2	0	0	5	773.3	1.4	0	0	0	7.8	0	0	0
0.6	0.9	1.2	8.10	0	4	1322.8	65.1	0.4	3.8	24.2	8.8	9.8	23.6	55.2
0.6	0.9	1.5	0	0	5	35.1	2.4	0	0	0	3.8	1.8	0.4	1.0
0.6	0.9	2	0	0	5	37.2	1.1	0	0	0	6.2	0	0	0
0.6	0.95	1.2	762.66	234.27	2	2938.7	197.6	0.2	1.4	15.0	9.2	49.2	78	75.8
0.6	0.95	1.5	79.20	37.36	3	2901.2	39.2	0.4	2.0	11.8	17.0	14.0	21.0	41.0
0.6	0.95	2	29.64	15.56	3	2080.2	3.4	0	0	0	15.8	0	0	0
0.9	0.9	1.5	90.02	90.02	5	28.3	6.5	0	0	1.0	7.4	1.8	2.4	6.2
0.9	0.9	2	62.32	62.32	5	45.1	1.6	0	0	0	8.8	0	0	0
0.9	0.95	1.5	1133.94	1133.94	5	39.1	14.1	0	0	1.6	5.0	1.0	2.4	9.6
0.9	0.95	2	1006.68	1006.68	5	72.1	2.1	0	0	0	11.2	0	0	0
Avg. / Total			208.61*	161.26	67	978.7	33.0	0.2	0.9	5.4	8.4	7.5	14.5	18.8

*The average of the upper bound refers to 79 instances to enable the comparison between algorithms.

Balancing Under No Sequencing Control   5

In this chapter, the assembly-line balancing is optimized for a production line under a random sequence of products. In contrast to Chapter 4, in which the models can be scheduled at wish, no control over the production sequence is assumed in the project presented in this chapter.

5.1   Problem Definition

The production system characteristics and the necessary assumptions to model an assembly line under random sequences are introduced in this section.

In contrast to the problem analyzed in Chapter 4, no active decision power about the order in which products are produced is assumed in this chapter. This assumption is not realistic for the operation of assembly lines in the automotive industry, however, it is a feasible alternative for the planning of an assembly line. The construction of an assembly line is usually performed without the knowledge of the future production sequence. By considering a random sequence, the planner opts for a pessimistic view of the product sequencing. Furthermore, in real applications, the assembly line planner does not has full control over the production sequence. Not only the final assembly line is responsible for the sequencing objective function, but also the body-in-white and painting sectors need to be considered in the solution. The optimal sequences for the multiple departments can differ. A second difficulty in integrating sequencing and balancing lays in the different time frames of these problems. The balancing problem is a strategic decision that happens in the planning phase of an assembly line, while the sequencing is a tactical or even operational decision that may be solved daily or weekly. Therefore, it is common that the balancing solution procedures do not consider the sequencing, aim at smoothening the workload at the balancing solution or use a proxy objective function [Emde et al.,

© The Author(s), under exclusive license to Springer Fachmedien Wiesbaden GmbH,   97
part of Springer Nature 2022
C. G. Stall Sikora, *Assembly-Line Balancing under Demand Uncertainty*,
Gabler Theses, https://doi.org/10.1007/978-3-658-36282-9_5

2010]. An extra justification for the lack of sequencing lays in the just-in-sequence production often employed in the automotive industry [Bukchin et al., 2002]. As the products can be highly customizable, the order in which they are ordered affects greatly the production due dates. Therefore, the production sequence is highly correlated with the apparently random order in which products are sold to the customers. Based on these issues of the products' sequencing, a random production sequence for the balancing of an assembly line is assumed here.

Considering a random inflow of products can provide algorithmic advantages since the sequencing problem does not need to be solved. The solved instances from Chapter 4 (Sikora [2021]), for instance, are limited to cyclical sequences of ten products. Under some assumptions, which are presented in this section, the balancing of lines with over millions or billions of product combinations is possible [Boysen et al., 2009a]. In the next paragraphs, the definition of a product model, as well as the production system characteristics and assumptions are described.

The definition of a model depends on the concept of tasks and options. A task is every indivisible operation that must be performed in the assembly of a product. Some tasks may be performed in different manners. For instance, the sunroof of a car may come in different forms and can be either manually or electrically controlled. These multiple varieties of a task are called *options* [Boysen et al., 2009b]. Each task (e.g., mounting the sunroof) is defined as a set of options (e.g., normal/wide, manual/electric) which may require different processing times. Tasks may present only a single option when the operation is common for every product, or multiple options, including an option that does not require any processing time (e.g., a car without a sunroof). A (product) model is defined as the combination of all the options necessary to assemble the product. These model combinations can account for astronomical numbers of possible products. The variety of an assembly line with 20 tasks and 2 options per task already surpasses a billion different product models.

The focus of this chapter lays on the optimization of paced assembly lines under a random production sequence. The objective of the optimization is to find the best assignment of tasks to stations in terms of operating costs. These costs are measured with respect to the station length and the expected utility work for the line's operation. An example for the realization of a production sequence is given in Figure 5.1 for one station, which is similar to Figure 4.1 in page 67. Again, the space between the left and the right line represents the station length, in which the operations are performed. The equivalent distance of a cycle time is marked with the vertical dashed line. Each horizontal bar represents a product, while the y-axis represents the time axis in the form of discrete cycles. The workers perform the tasks until a product is finished, and then return to the beginning of the line to start the operations for the next product. If the next product is not yet available (as is the case of $P3$ after finishing $P2$), idle time occurs, as the worker must wait for the next piece. If a product requires more than the cycle time (as in $P3$), the worker may continue the operations even though the next piece ($P4$) already enters the station. The start position of the next piece ($P4$), however, is not at zero in this case. If at any moment the operation cannot be finished within the station's boundary, a utility worker may be called to aid with the operations. The amount of time used by the utility worker is called "utility work", which is a proxy for the variable operational cost of the line. In the figure, the gray striped bar is used to denote the utility work that is used to avoid the boundary violation shown by the dashed bar.

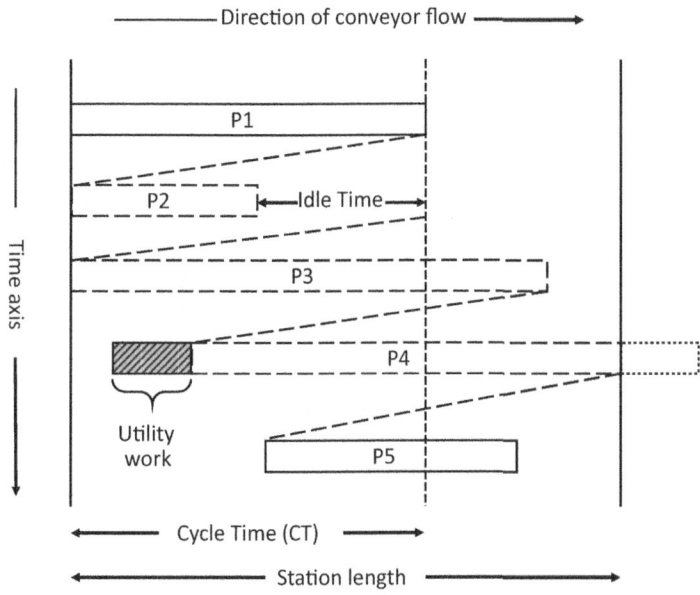

Figure 5.1 Realization of a production sequence in a paced assembly line. Utility work is used to assure processing within station boundaries

The worker schedule for the given product sequence as presented in Figure 5.1 is proven to be optimal by Yano and Rachamadugu [1991]. The simple scheduling optimality rule states that the worker performs the operations of a model until its completion before he or she moves to the next product piece. Utility work is only used when the regular worker would otherwise not be able to finish the operation within station boundaries.

There is a trade-off between the station length and the station's expected amount of utility work. The same sequence realization is shown alternatively in Figure 5.2 for a station with a longer length. Due to the extra flexibility, models $P1$ to $P5$ can be processed by the regular worker without the need of utility work.

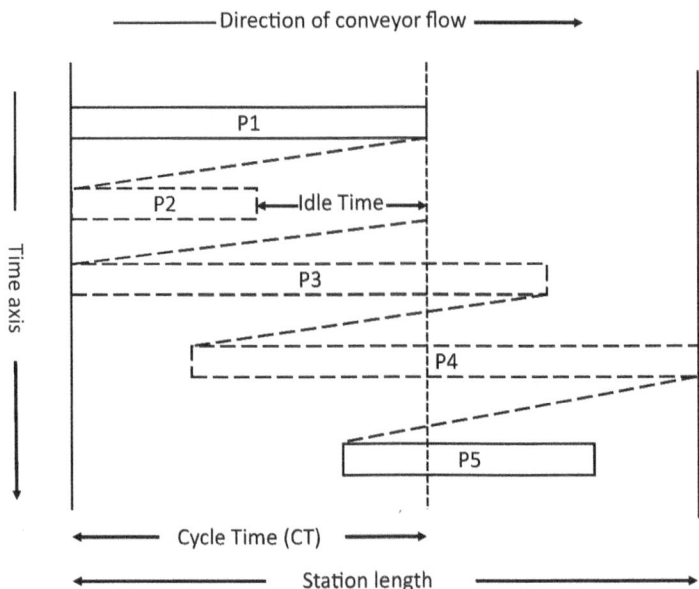

Figure 5.2 Comparison of a realization of a production sequence in a paced assembly line with longer station length. No utility work for P1–P5 is needed

As illustrated in Figure 5.1, the station borders are fixed and cannot be exceeded. This restriction is important so that there is no interference between the workers of adjacent stations. Although all workers are bound by the same sequence of products, their internal scheduling is independent. This independence is a critical assumption for the solution method since the problem can in this case be decomposed. Although the line can accommodate a huge number of possible models, each station processes a limited number of tasks. The number of different possible combinations within a station is rather limited. If 5 tasks with two options each are performed in a station, locally there exist only 32 different models, which is much less than the total number of combinations for the whole line. In order to treat only the locally observable variants within a station, another assumption is required: the task processing times are independent as well. That is, the processing time of each task does not depend on the options selected for the other tasks.

Based on the problem description and the assumptions, the problem is defined as the assignment of tasks T to a given set of stations S while minimizing the operational costs. Each task t has a set of options $o \in O_t$ with different integer processing times Dur_{to} and a given relative demand Dem_{to}. The average processing time for each task is expressed as Dur_t^{avg}. To consider integer processing times, the real task duration can be multiplied by 10, 100, or 1000 and rounded to the nearest integer, which is common in the assembly-line balancing literature [Scholl et al., 1999]. Assignments must obey precedence relations contained in the paired set $Prec$. The number of stations is considered to be externally given so that the variable station cost consists of the length of the stations and the expected utility work. It is sensible to assume that the cost of the station length grows linearly with the length and is related to the cost of the investment and maintenance of the conveyor belt and the opportunity cost of the plant space. The cost of the utility work is assumed to be a linear function of the expected utility work, which is described in more detail in Section 5.2. The optimization model is then given by

$$\textbf{Minimize } c_1 \cdot \sum_{s \in S} Len_s + c_2 \cdot \sum_{s \in S} E\left(C(x_{1s}, ..., x_{|T|s}, Len_s)\right) \tag{5.1}$$

$$\sum_{s \in S} x_{ts} = 1 \quad \forall\, t \in T \tag{5.2}$$

$$\sum_{k \in S: k \leq s} x_{t_1 k} \geq \sum_{k \in S: k \leq s} x_{t_2 k} \quad \forall\, (t_1, t_2) \in Prec, s \in S \tag{5.3}$$

$$\sum_{t \in T} Dur_t^{avg} \cdot x_{ts} \leq CT \quad \forall\, s \in S \tag{5.4}$$

$$x_{ts} \in \{0, 1\}, Len_s \geq 0 \quad \forall\, t \in T, s \in S. \tag{5.5}$$

There are two decision variables, the assignment variables x_{ts} of a task t to a station s and the length Len_s of a station s. The objective function is described in expression (5.1). The total cost containing the weighted sum line length costs and the expected utility work of an assignment $\sum_{s \in S} E\left(C(x_{1s}, ..., x_{|T|s}, Len_s)\right)$ under a random sequence of products is minimized. C is the amount of utility work required based on an assignment and a station length. The calculation of the expected utility work expressed in the objective function (equation (5.1)) is discussed in Section 5.2. Equation (5.2) is the occurrence restriction. The precedence constraints are modeled in expression (5.3). Expression (5.4) is a restriction on the average processing time in each station. Utility workers can be used to compensate sequences of heavily loaded products, but the average processing time must be within the cycle time (CT), otherwise, the assembly line is undersized. Finally, the assignment variables x_{ts} are binary, while the station length Len_s is non-negative (restriction 5.5).

5.2   Evaluating the Expected Utility Work

In this section, it is shown how to obtain the expected utility work of a station for a given task assignment and station length. The utility work calculation is based on the different varieties of products that can be processed in the station. As it is assumed that the tasks are independent of each other, only the combinations of options of the tasks assigned to the given station must be considered. The formulation as a Markov process is inspired by the applications on balancing and sequencing of unpaced lines by Lopes, Sikora, Michels, Lindbeck da Silva and Magatão [2018] and sequencing with stochastic release dates by Gwiggner [2020].

The set of models observable in a station s is named M here. A model m is defined by the subset of options selected for the product. From each task t assigned to the station s, exactly one option out of the possible multiple options must be performed. The set of the selected options for model m is given by the set O_m. The processing time of model m in station s is then given by

$$Dur_{ms} = \sum_{t \in T, o \in O_m} Dur_{to} \cdot x_{ts} \ \forall \ s \in S, m \in M.$$

Each model—as seen from the perspective of a station s—also has a probability P_{ms} of being ordered. Based on the individual probability P_{to} of one option o to be purchased, the probability of a given model m as seen from station s is given by

$$P_{ms} = \prod_{o \in O_m, t \in T | x_{ts} = 1} P_{to} \ \forall \ s \in S, m \in M.$$

For this work, the options are assumed to be independent of each other and therefore the product of the individual probabilities can be used. The evaluation using Markov chains is, however, not limited to independent purchasing probabilities. Correlations could also be used in the calculation of the relative demand (P_{ms}) of each model m. For simplicity, only the independent case is considered further. Furthermore, as the formulation is decomposed based on stations, the station index s is dropped in the notation for the rest of this section since the calculation can be applied for each station individually.

The Markov process used to calculate the expected utility work is based on the position of the worker after finishing a piece. As it can be seen in Figure 5.1, the worker advances in the line while processing a model and then returns to the beginning part of the station to start processing the next model. This return can account for up to CT equivalent length units (the launching interval between two products). Considering a worker position Pos before the processing of a product with processing time Dur_m, the new initial position of the worker in the next cycle is given by $Pos + Dur_m - CT$ (assuming no utility work and no idle time has occurred). The relative displacement ($Dur_m - CT$) is labeled as Δ_m and is illustrated in Figure 5.3.

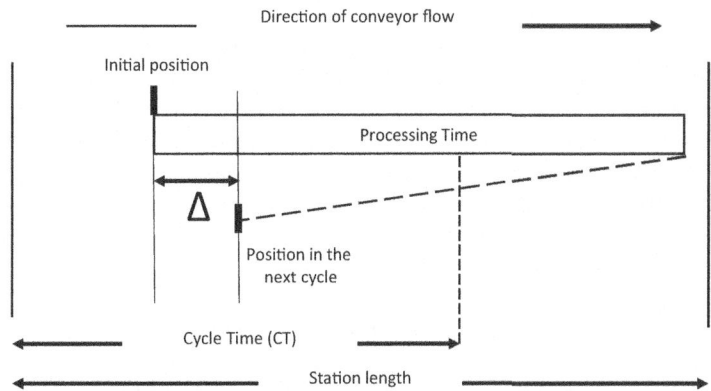

Figure 5.3 Relative displacement of the worker

The worker's initial position (Pos) has to respect some boundaries. The lower bound is given by the station start, since the products cannot be accessed before the station. A upper bound is given by the station length minus the cycle time. This upper bound is achieved when the worker finishes the operations at the end of the line and returns CT units, resulting in the interval [0; $Len - CT$].

Based on the displacement Δ_m of each model and the position Pos of a worker, the next position Pos^q is calculated as $Pos^q = \max(0, \min(Len - CT, Pos + \Delta_m))$. A Markov process can be defined in which each state is a possible starting worker position within the interval [0, $Len - CT$] and the next position is given by Δ_m. A transition matrix Tr consists of transition probabilities between the multiple states. The entries of the matrix are the probabilities that a random model has a specific Δ_m value. In order to express the matrix in a compact form, $\overline{Len} = Len - CT$ is used in the definition of Tr. The transition matrix of such a Markov process is expressed by

$$
Tr = \begin{array}{c} \\ 0 \\ \vdots \\ i \\ \vdots \\ \overline{Len} \end{array}
\begin{array}{ccccc}
0 & \cdots & j & \cdots & \overline{Len} \\
\left(\begin{array}{ccccc}
P(\Delta \leq 0) & \cdots & P(\Delta = j) & \cdots & P(\Delta \geq \overline{Len}) \\
\vdots & \ddots & \vdots & \ddots & \vdots \\
P(\Delta \leq -i) & \cdots & P(\Delta = j - i) & \cdots & P(\Delta \geq \overline{Len} - i) \\
\vdots & \ddots & \vdots & \ddots & \vdots \\
P(\Delta \leq -\overline{Len}) & \cdots & P(\Delta = j - \overline{Len}) & \cdots & P(\Delta \geq 0)
\end{array} \right)
\end{array}
$$

where rows represent the start position of of the worker and columns the worker position after processing the piece (or the initial position for the next product). The value of Δ represents the displacement and links both the initial and following state. The entries in the matrix are the correspondent probabilities that products with a given Δ are processed in the line. The matrix assures that the position always lies in the feasible interval, either by assuming idle time or utility work. The necessary utility work to not exceed the station boundary for a given model m and position Pos is given by

$$U(\Delta_m, Pos) = \text{Max}(0, Pos + \Delta_m - (Len - CT)). \tag{5.6}$$

The transition matrix Tr can be used to calculate the stationary probabilities w_{Pos} of a worker being in a given position $Pos \in [0, Len - CT]$ by solving the system of linear equations $w = Tr \cdot w$ and $\sum_{Pos \in [0, Len - CT]} w_{Pos} = 1$ [Cechin and Corso, 2019]. The stationary probabilities combined with the probability P_m of each model product can then be used to calculate the expected utility work for a given assignment and station length using the expression

$$E\left(C(x_{1s}, ..., x_{|T|s}, Len_s)\right) = \sum_{m \in M} \sum_{Pos \in [0, Len - CT]} P_m \cdot w_{Pos} \cdot U(\Delta_m, Pos). \tag{5.7}$$

5.3   Properties of the Total Cost Function

This section contains the proof of two important properties of the total cost function: the cost function is i) piece-wise linear and ii) convex for the assumptions presented in the problem description. While the station-length cost is linear, the expected utility-work function is evaluated in this section.

Theorem 1 *The expected- utility-work function is piecewise linear in the station length.*

Proof. Suppose the same station is analyzed in two different layouts with length n and $n + \delta$, in which n is a natural number and $0 < \delta \leq 1$. Consider that the worker of the station is in the same starting position i, while the product sequence and all processing times are identical. The position of the worker lies in either one of the two configurations shown in Figure 5.4. From an initial position i_0, the worker has the same position in both cases if the right border is not surpassed.

A transition of configuration 1 to configuration 2 occurs when the final position of the worker would be larger than n. In this case, utility work occurs. As assumed in the problem description, all processing times have integer values. If a product requires $U_1 \geq 1$ units of utility work in the station of length n from a line in configuration 1, the station with length $n + \delta$ requires $U_1 - \delta$ units of utility work for the same operation.

In configuration 2, every sequence of products requires exactly the same amount of utility work, since the distance of the worker position and the end of the station is identical.

When idle time occurs, a transition from configuration 2 to configuration 1 happens. As all processing times are integer, any amount of idle time would result in a transition to configuration 1.

Summing up all the effects, both layouts do not need utility work if the system stays in configuration 1, the same amount of utility work is needed in configuration 2, and no utility work again in the transition from configuration 2 to 1. In each transition from configuration 1 to 2, U_1 and $U_1 - \delta$ units of utility work are needed, respectively.

The probabilities for each transition are identical for both layouts of the station—independent of the value of δ. Hence, the expected utility work varies linearly with the value of δ. Therefore, the expected utility-work function is piecewise linear. $\square$

Theorem 2 *The expected-utility-work function is convex.*

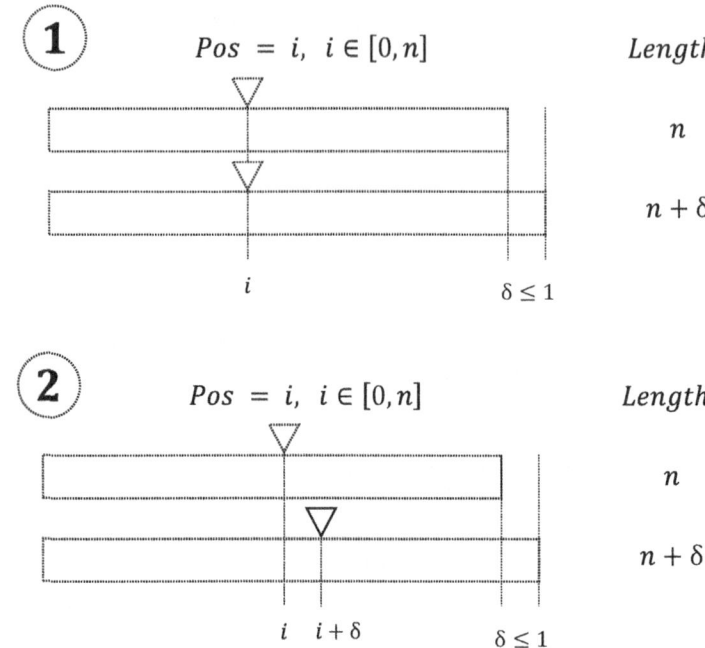

Figure 5.4 The two possible configurations of a station with lengths n and $n+\delta$, for $0 < \delta \le 1$

Proof. The convexity proof consists of two parts:

i) the expected utility work function decreases with respect to the length.

ii) $U_{n-1} - U_n \ge U_n - U_{n+1}$, where U_n refers to the expected utility work of a station with length n.

Part i): see proof of Theorem 1 for $\delta \in (0, 1]$.

Part ii): Consider three layouts of a station with length $n - 1$, n, and $n + 1$ with a worker at the starting position $i_0 \in [0, n - CT - 1]$. For the same sequence of products, the position of the worker after finishing the tasks of a product and return CT equivalent units to the left is given by one of the four configurations displayed in Figure 5.5.

The four possible configurations are:

- Configuration 1: the worker is at the same position i (for $i \in [0, n - CT - 1]$) in all layouts of the station.
- Configuration 2: the worker is at position i for the layouts with length n and $n + 1$, while the worker of the layout with length $n - 1$ is at position $i - 1$ (for $i \in [1, n - CT]$).
- Configuration 3: the positions of the worker in the layout with length $n - 1$, n, and $n + 1$ are, respectively, at $i - 1, i, i + 1$ (for $i \in [1, n - CT]$)
- Configuration 4: the worker in the layout with length $n - 1$ and n are in position i, while the worker in the layout with length $n + 1$ is at position $i + 1$ (for $i \in [0, n - CT - 1]$).

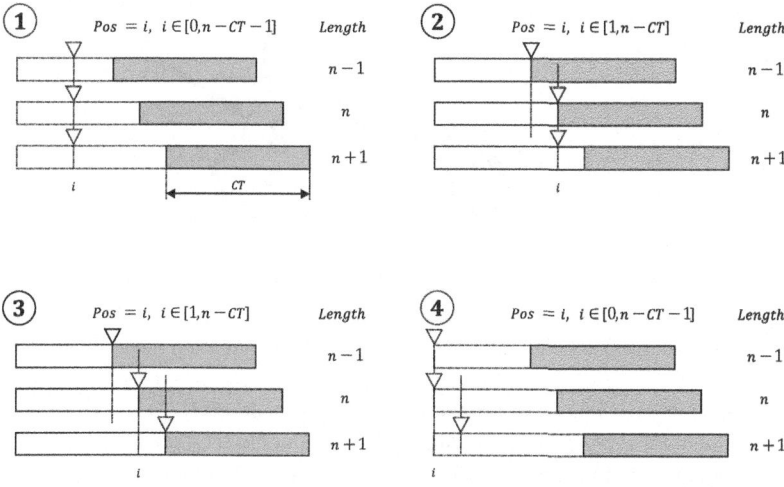

Figure 5.5 The four possible position configurations after a sequence q of products for stations with length $n - 1$, n, and $n + 1$

Other configurations may be possible at the start of the operation with arbitrary positions for the three station layouts. During the operation, however, the arbitrary configuration will approach configurations 1 or 4 if idle time occurs or configurations 2 or 3 if utility work occurs. Because of this, only the four presented configurations are said to be stable in a steady-state production.

Figure 5.6 describes all the possible transitions between the stable configurations. If the position of the worker in all cases does not exceed any station boundary, the configuration remains unchanged. From configuration 1, if the position of the worker of the station with length $n - 1$ would be n, exactly one unit of utility work occurs. The resulting position is in configuration 2. If at least one unit of utility work occurs in stations of length $n - 1$ and n, the relative position of the workers must be as in configuration 3. Finally, when idle time occurs, configuration 2 transitions to configuration 1, and configuration 3 transitions to configuration 4 if the station of length $n - 1$ has exactly one unit of idle time or to configuration 1 if more idle time occurs. No other transition is possible.

Figure 5.6 The possible transitions between the four possible position configurations in three layouts of a station with lengths $n - 1$, n, and $n + 1$

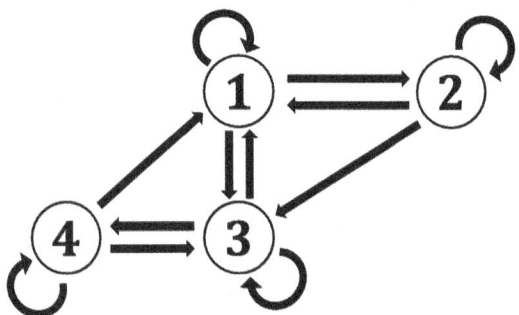

The amount of utility work for all possible transitions is described in Table 5.1. Some transitions result in 0, 1, or an arbitrary amount of utility work. The arbitrary amounts are described with U_{ij} for the transition of configuration i to configuration j.

In order to prove that $U_{n-1} - U_n \geq U_n - U_{n+1}$, the value of $U_{n-1} + U_{n+1} - 2 \cdot U_n$ is calculated for every possible transition. The net amount $(U_{n-1} + U_{n+1} - 2 \cdot U_n)$ of the utility work in every transition is displayed in the last column of Table 5.1. Given probabilities P_{ij}^T of a transition of configuration i to j, the expression results in

$$U_{n-1} + U_{n+1} - 2 \cdot U_n = +1 \cdot P_{12}^T - 1 \cdot P_{23}^T + 1 \cdot P_{43}^T = P_{12}^T - P_{23}^T + P_{43}^T.$$

From Figure 5.6, it can be observed that configuration 2 can only be reached from configuration 1 (Transition $1 \to 2$) and configuration 2. Therefore, the frequency of transition $1 \to 2$ is higher or equal to transition $2 \to 3$, since the transition $2 \to 1$ is also possible. As the average of the expected processing time of each station must be lower than or equal to the cycle time, the probability of idle time (P_{21}^T) is non-zero (assuming that there is at least a model whose processing time is longer than the cycle time). Therefore, $P_{12}^T > P_{23}^T$.

Table 5.1 Utility work necessary for each transition for three layouts of a station with lengths $n-1$, n, and $n+1$

Transition	Length			$U_{n-1}+U_{n+1}-2\cdot U_n$
	$n-1$	n	$n+1$	
	Utility work			
$1 \to 1$	0	0	0	0
$1 \to 2$	1	0	0	+1
$1 \to 3$	$U_{13}+1$	U_{13}	$U_{13}-1$	0
$2 \to 1$	0	0	0	0
$2 \to 2$	0	0	0	0
$2 \to 3$	U_{23}	U_{23}	$U_{23}-1$	-1
$3 \to 1$	0	0	0	0
$3 \to 3$	U_{33}	U_{33}	U_{33}	0
$3 \to 4$	0	0	0	0
$4 \to 1$	0	0	0	0
$4 \to 3$	$U_{43}+1$	U_{43}	U_{43}	+1
$4 \to 4$	0	0	0	0

Inequality $P_{12}^T > P_{23}^T$ results in $P_{12}^T - P_{23}^T + P_{43}^T > 0$, so that $U_{n-1} + U_{n+1} - 2 \cdot U_n > 0$ and, consequently, $U_{n-1} - U_n > U_n - U_{n+1}$.

For the case in which no model is longer than the cycle time, no utility work is necessary and $U_{n-1} = U_n = U_{n+1} = 0$, which is also convex. ☐

5.4 Solution Algorithm

The proposed algorithm for the balancing of assembly lines for random sequences is based on a Branch-and-Bound algorithm. The structure of the procedure is illustrated in Figure 5.7. Nodes consist of a set of tasks, whose enumeration is described in section 5.4.1. For each node, the station length is optimized using an exponential fitting algorithm described in section 5.4.2. For each combination of task assignment and station length, a Markov chain (section 5.2) is used to determine the expected utility work and consequently the assignment cost. The exact algorithm is described in section 5.4.3.

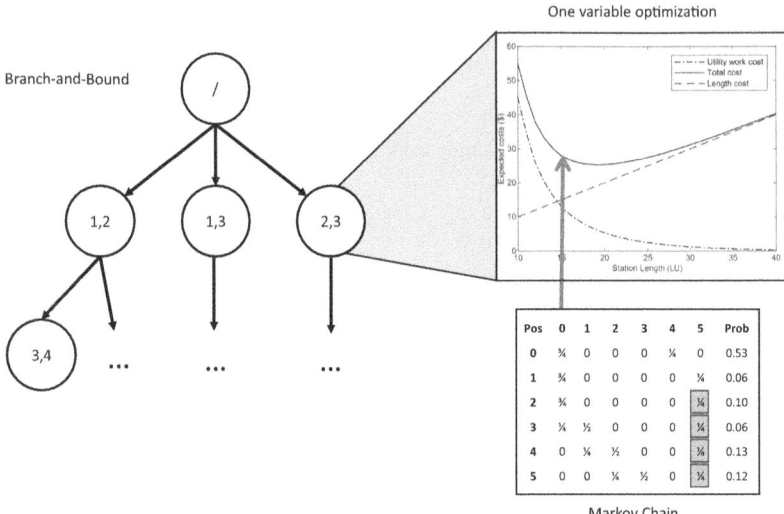

Figure 5.7 Representation of the algorithm structure

5.4.1   Node Enumeration Scheme

The search for the best assignment is performed in a search tree, in which nodes represent feasible assignments of tasks. A solution of the model of section 5.1 is considered feasible if all tasks are assigned, task assignments respect precedence relations, and stations' expected processing times do not exceed the cycle time. As only the expected processing time is relevant for the feasibility, the solution space of the stochastic problem is identical to a deterministic problem considering the expected processing times. Although the assessment of the objective function value in the stochastic case is more costly, the enumeration of feasible solutions can use known procedures for the feasibility version of the simple assembly line balancing problem (SALBP-F).

Among the branching schemes for SALBP, the best-performing methods are based on the station-oriented enumeration schemes [Scholl and Klein, 1997; Sewell and Jacobson, 2012]. In this scheme, each node of the tree consists of a set of tasks assigned to a station. Each level k of the enumeration tree consists of the nodes of possible assignments to station k. In the proposed algorithm, the search starts with a root node at level 0 with no tasks assigned. The enumeration is performed station-wise along with the stations of the line.

Algorithm 5.1 is used to enumerate the assignment alternatives of a station for a set of available tasks L. This set contains the tasks whose predecessors are already assigned so that their assignments respect the precedence constraints. Algorithm 5.1 performs a recursive deep-first search to enumerate all the possible assignments. The algorithm is an adaptation of the enumeration heuristic of Hoffmann [1963] and Fleszar and Hindi [2003]. In their heuristic, only the best assignment of each level is chosen. In the proposed branch-and-bound, all feasible nodes are enumerated and stored.

During the node exploration, some feasibility rules are checked. The precedence relations are obeyed in the construction of the set L. Whenever a task is added or removed to the set of assigned tasks A, set L is updated. That is, elements are added or removed from L so that all tasks in L can be directly assigned next. The algorithm is initiated in each level with $q = 1$ so that the index r starts with the first element of set L (L_1). Set L at the root node of the algorithm contains the tasks without predecessors.

The second restriction is the expected processing time, which must be less than or equal to the cycle time in each station. This restriction is equivalent to the balancing of a single model with the average processing times and is implemented in the first if-clause in Algorithm 5.1. In the enumeration process, it is important to track the idle time of the assignments. The idle time for the average model of a station is the difference between the cycle time and the average processing time in the station. For the whole line, the idle time of the average model is given by

$$CT \cdot |S| - \sum_{t \in T} Dur_t^{avg}.$$

This quantity is distributed among the stations since not all assignments fill up the station up to the cycle time. In the enumeration process, assignments that use more than the available idle time are infeasible, since it is impossible to assign the remaining tasks of the average model to the remaining stations. Therefore, at the generation of a node, the remaining idle time is used as a feasibility condition. This condition is checked in the second if-clause.

For the deterministic SALBP, Jackson [1956] presented the maximum load rule for station assignments. This dominance rule states that assignments with non-maximal station load are dominated by maximum loads for the SALBP-1 variant. In the stochastic case, this rule does not apply, because the expected processing time may require much more utility work than assignments with positive expected idle time. Therefore, Algorithm 5.1 uses a proposed weaker version of this dominance rule: the maximum upper load rule. This rule considers the longest processing times of each task t (Dur_t^{max}) and does not enumerate assignments whose processing time in the worst-case scenario is small enough that another task could be added without causing utility work. This rule is implemented in the third if-clause. This restriction is waived for the last station of the enumeration process since set L is empty.

Algorithm 5.1: OnePackingSearch(q, $IdleTime$, L, A)

---

for r: q to $|L|$ **do**
  $i = L_r$
  if $\sum_{t \in A} Dur_t^{avg} + Dur_i^{avg} \leq CT$ **then**
    $A = A \cup \{i\}$
    Update L
    if $CT - \sum_{t \in A} Dur_t^{avg} \geq IdleTime$ **then**
      if $\sum_{t \in A} Dur_t^{max} + min_{k \in [i, |L|]}(Dur^{max}(L_k)) > CT$ **then**
        $RemainingIdleTime = IdleTime - (CT - \sum_{t \in A} Dur_t^{avg})$
        CreateNode(A, $RemainingIdleTime$, $L \setminus A$)
      OnePackingSearch($r + 1$, $IdleTime$, L, A)
      $A = A \setminus \{i\}$
      Update L

In Algorithm 5.1, function 'OnePackingSearch' is responsible for the enumeration of the assignments. For the feasible station loads, function 'CreateNode' stores the given assignment, the remaining idle time for the next stations, and the list of tasks that may be assigned next ($L \setminus A$).

5.4.2 Length Optimization

Each assignment or node of the enumeration tree has a cost compounded by the length cost and the expected utility work cost of this station. The expected utility work is calculated by a Markov chain presented in Section 5.2 and depends on the station length, as illustrated in an example in Figure 5.8. The expected utility-work cost is a convex and piece-wise linear non increasing function (proof in Section 5.3) while the line length cost is a linear function. The total cost function, as in the objective function of the model (expression 5.1), is the weighted sum of both costs with weights c_1 and c_2 which represent monetary units per length unit and monetary units per time unit, respectively. An example of the total cost curve is given in Figure 5.9.

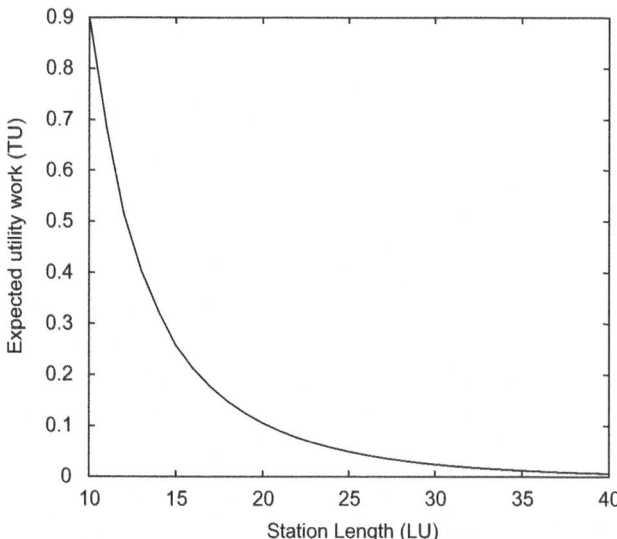

Figure 5.8 Expected utility work with respect to station length for an example. Although the curve resembles an exponential curve, it is a piecewise linear function

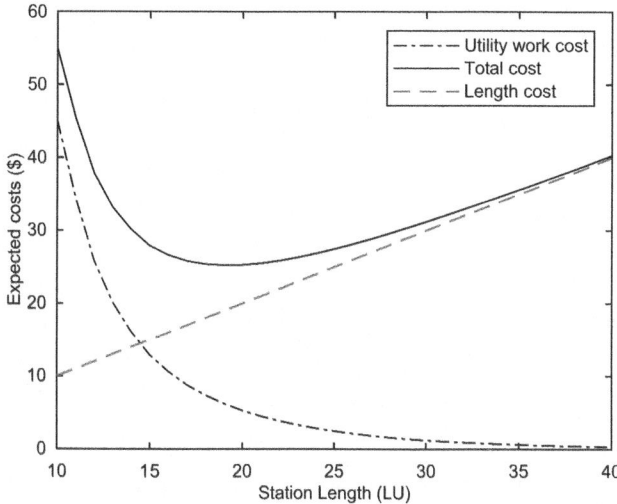

Figure 5.9 Expected utility work cost, line length cost, and expected total cost with respect to station length for an example

In order to obtain the minimal cost of a given assignment, the optimal length of the station must be determined. In this section, a one-variable optimization algorithm is presented. The processing-time data is considered integer (and as a consequence all Δ-values are integer) so that the optimal length is also integer since the total cost function is piecewise linear with integer nodes. The optimal station length can be obtained by generating the corresponding Markov chain for each station length and choosing the minimal cost value. In order not to enumerate all possible station lengths, a search algorithm is proposed.

The structure of the one-variable optimization is shown in Algorithm 2. As the total cost function is convex, but no information of the derivatives is available, the procedure performs the evaluation of a length value per iteration and uses the information of other three known length values to determine the new search interval. These three values correspond to a lower bound, an upper bound, and a third known point, which is called $Pivot$. The algorithm structure is based on search procedures such as the Fibonacci Search [Ferguson, 1960] or the Golden Section Search [Kiefer, 1953]. In contrast to methods for arbitrary curves, the proposed algorithm uses the information of the shape of the curve to select the new evaluated point. As the expected utility-work curve resembles an exponential shape, an exponential fitting is used to approximate the optimal solution and accelerate the search procedure. Algorithm 2 also contains a correcting function for the cases in which the exponential fitting provides approximations outside the search interval. Finally, the search interval is updated based on the new evaluated search length.

Algorithm 5.2: LengthOptIteration(A, LB, $Pivot$, UB, TC_{LB}, TC_{Pivot}, TC_{UB})

```
/* Approximate total cost curve and calculate next
   point.                                           */
```
$Sol = $ ExponentialFitting(LB, $Pivot$, UB, TC_{LB}, TC_{Pivot}, TC_{UB})
```
/* Assure Sol is within known bounds.              */
```
CorrectWithinBounds(Sol, LB, $Pivot$, UB)
```
/* Calculate the expected utility work.            */
```
$TC_{Sol} = c_1 \cdot Sol + c_2 \cdot$ MarkovChain(A, Sol)
```
/* Uptade bounds.                                  */
```
UpdateBounds(Sol, LB, $Pivot$, UB, TC_{Sol}, TC_{LB}, TC_{Pivot}, TC_{UB})

In the "ExponentialFitting" function ($\tilde{E}(Len)$), the known points ($Pivot$, LB, and UB) and their respective total-costs (TC_{Pivot}, TC_{LB}, and TC_{UB}) are used to approximate the total-cost curve by a compound exponential and linear curve in form

$$\tilde{E}(Len) = c_1 \cdot Len + c_2 \cdot a \cdot e^{-b \cdot Len}.$$

Among many forms to fit the curve for the given data, the version of the algorithm that performed best in the empirical tests uses only two points (P_1 and P_2) for the fitting. These points are the pair of length points among the three explored in the iteration that are closer to each other. Therefore, one of the points is $Pivot$ and the second is either the lower bound (LB) or upper bound (UB), whichever has the lowest distance to $Pivot$. The two-point fitting results in parameters

$$b = \ln\left(\frac{TC_{P_1} - c_1 \cdot P_1}{TC_{P_2} - c_1 \cdot P_2}\right) / (P_2 - P_1)$$

and

$$a = \frac{(TC_{P_1} - c_1 \cdot P_1) \cdot e^{b \cdot P_1}}{c_2}.$$

The optimal solution can then be found by

$$\frac{d\tilde{E}(Len)}{dLen} = 0$$

and is given by

$$Sol = \left(\ln(c_2/c_1) + \ln a + \ln b\right) / b.$$

Although the fitting of the utility work with an exponential curve provides a good approximation, the algorithm does not always converge to the optimal solution since the curve is not an exponential function. Therefore, the "CorrectWithinBounds" function is used to make sure the new test point is within the search interval. This function does not alter the value of Sol if its value lays within the interval $[LB, UB]$ and is different from $Pivot$. In the case of a bound violation (either $Sol > UB$ or $Sol < LB$), Sol is corrected to the middle point of $Pivot$ and the violated bound rounded down. If the selected interval is of length 1 (for instance $Pivot = LB + 1$), Sol is corrected as the middle point of the other interval (in the example, the middle point between $Pivot$ and UB). Finally, if the approximation gives $Sol = Pivot$, then Sol is set to the middle point of the larger of the intervals $[LB, Pivot]$ or $[Pivot, UB]$. This way, Sol is a non-evaluated point within the search interval at the beginning of each iteration. The calculation of the total cost of Sol uses the "MarkovChain" function described in Sec. 5.2.

The remaining step in the procedure is given by the function "UpdateBounds". In this procedure, the search interval is updated based on the evaluation of the new point *Sol*. The procedure is illustrated in Figure 5.10. From the four known points, three are stored for the next iteration. Either the lower or upper bound is updated with one of the internal points (*Sol* or *Pivot*). The case in which the left bound is updated is shown in Figure 5.10a, while the upper bound update is depicted in Figure 5.10b.

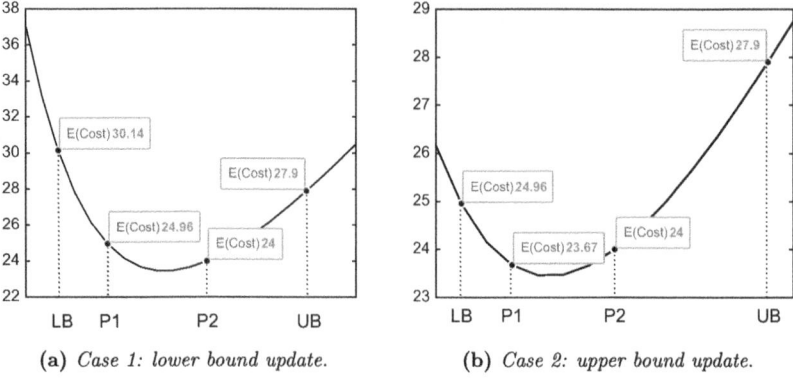

(a) *Case 1: lower bound update.*   (b) *Case 2: upper bound update.*

Figure 5.10 Illustrated cases for the bounds update function

Algorithm 2 requires as initialization the value of three points (LB, UB, and $Pivot$). As a lower bound for the length, the cycle time (CT) equivalent is used, while the first $Pivot$-value is $CT + 1$. The utility work evaluation for $Len = CT$ requires only the weighted sum of the processing times longer than the cycle time since there is no length flexibility. For $Len = CT + 1$, the corresponding Markov chain has only two states (Position 0 and 1) and is also fast to calculate. An initial upper bound for the length is given by

$$UB = \left\lfloor \frac{\min \left(TC_{LB}, TC_{Pivot} \right)}{c_1} \right\rfloor .$$

This bound exhibits the same length cost as the better total cost of the LB or $Pivot$ solution. As utility-work costs are not considered, UB must be an upper bound for the length of the station. To avoid calculating a large Markov chain, TC_{UB} is initialized with an infinite value, and UB is not used for the exponential fitting until an upper bound with a finite cost is calculated in the execution of the algorithm.

5.4.3 Iterative Computation of Nodes

With the node enumeration (Subsection 5.4.1) and node evaluation (Subsection 5.4.2) procedures defined, the last piece of the solution procedure is the search strategy.

Empirical tests show that the computational time required to evaluate a node is much larger than the time required to compute the feasibility of child nodes in the enumeration procedure. This is very unfortunate in the case of paths with feasible assignments for stations at the beginning of the line and infeasible nodes in deeper levels of the search tree. Therefore, the proposed search procedure generates the search tree node before the evaluation of the nodes.

Another drawback of the procedure is the lack of known lower bounds for a given assignment before the evaluation of the node. Since the utility work is the result of a Markov Process and length is also optimized for each node, it is difficult to infer a proper lower bound for the total cost just based on the balancing variables. If, however, a feasible interval for the optimal length $[LB, UB]$ for a given assignment is known, a lower bound can be calculated by

$$TC_{Node} \geq c_1 \cdot LB + (TC_{UB} - c_1 \cdot UB).$$

A valid lower bound is given by the length cost of the LB and the utility work cost of the UB (that is, total cost minus length cost). This lower bound is justified by the fact that length cost increases in length, while utility work cost decreases for increments of lengths.

Along with the node's lower bound on the total cost, a lower bound for the line cost can be defined for each node. The line cost lower bound evaluated in a node is the shortest path from the root node to a leaf node including the evaluated node. The cost lower bound of each node is used as a distance measure for the computation of the shortest path. In this chapter, a node lower bound (or upper bound) is shortened by 'Node LB (or UB)', and a line lower bound is denoted by 'Line LB'.

Based on the problem characteristics and the proposed lower bound on total cost, the proposed algorithm employs an iterative exploration of nodes. That is, the nodes are not solved until convergence initially. The algorithm loops between the feasible nodes and calculates one iteration of the 'LengthOptIteration' at a time for each node. The new interval is used to update lower bounds and potentially prune nodes in the process.

The algorithm is outlined in Figure 5.11. The procedure is described in 12 steps:

1. Load input data: cycle time, precedence relations, processing times of options, and probabilities of options.
2. Enumerate nodes: described in Subsection 5.4.1.
3. Set a path based on the cost approximation: to provide an upper bound, one path is selected heuristically. The first iteration of 'LengthOptIteration' (Subsection 5.4.3) is calculated. The result of the first exponential fitting is used as an approximation for the total cost. The path with the lowest approximated cost is selected as the first solution.
4. Solve nodes in the path: the nodes on the initial path are solved to convergence.
5. Line UB: the sum of the cost of the solved nodes in the path provides an upper bound on the line cost.
6. While nodes in the tree: The main loop of the procedure. The iterative exploration of nodes is performed until all nodes are pruned.
7. Select node: a node is selected for exploration. In the implementation, the nodes with the lowest line lower bound are selected. In computers with multiple cores, more than one node may be selected at the same time, since their computations are independent.
8. LengthOptIteration: an iteration of the one-variable optimization is performed for the selected node.
9. Node LB: actualize the cost lower bound of the node.
10. If clause: decides whether the node is solved. If the search interval $[LB, UB]$ has unexplored lengths, go directly to 11. If the one-variable search is completed, the node upper bound on cost is updated before starting step 11.
11. Update Tree UB/LB: new bounds on the selected node are propagated to the line cost bounds of parents and child nodes.
12. Remove nodes with $LB \geq UB$: if a newly actualized line cost lower bound is greater than or equal to the line cost upper bound for any node, this node is pruned. After the pruning, go back to step 6.

5.5   Tests and Results

5.5.1   Dataset

For the balancing of assembly lines under random sequences, a dataset is proposed based on the SALBP instances provided by Scholl [1993]. The selection of Scholl's dataset instead of the newer Otto et al. [2013] dataset is justified based on a larger variety of instance size. The structure dataset of Otto et al. [2013] contains instances

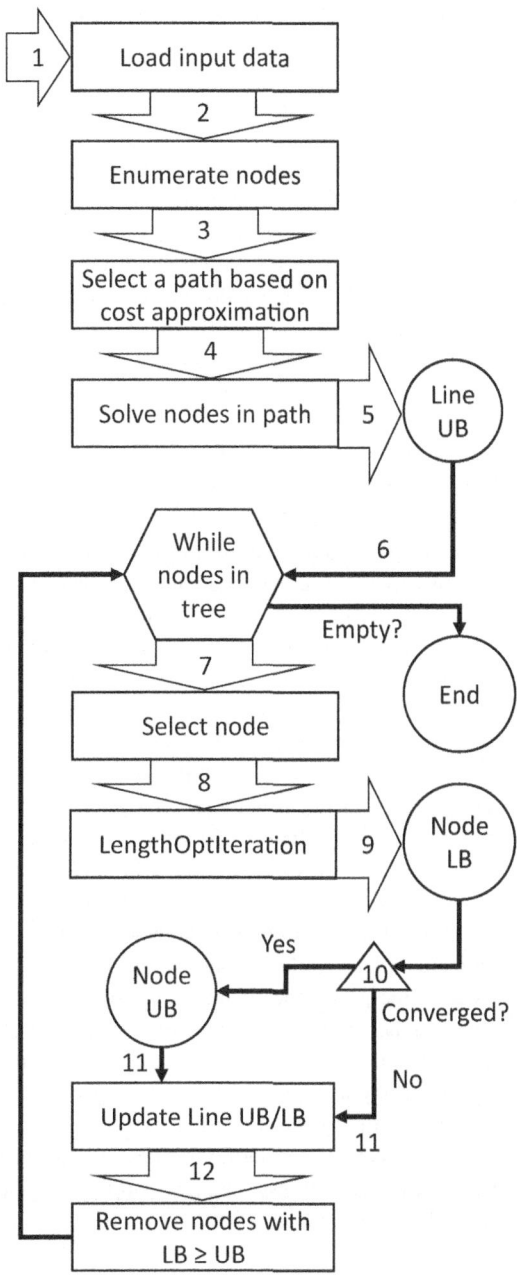

Figure 5.11 Flowchart of the proposed algorithm for assembly-line balancing under random sequences

with 20, 50, 100, and 1000 tasks. In preliminary tests, instances with 20 instances are easy to solve and several of the instances with 50 tasks are not solvable due to memory requirements using the proposed algorithm. Therefore, the instances from Scholl [1993] ranging from 8 to 297 tasks are used to generate a dataset for the problem.

For this problem, the tasks are modeled as options that can be added to the product. For the generated instances, each one of the tasks of an SALBP instance is generated randomly with 1, 2, or 3 options with the same probability. If a task contains 2 or more options, the second option has a 20% chance of having no processing time, that is, the option is not built into the product. For the other cases, a random integer value between 50% and 150% of the processing time of the SALBP task is drawn from a uniform distribution. For the relative probabilities of each option being purchased, random uniform values for each option are generated. The random values are then divided by their sum and rounded to decimal places so that the sum of the relative probabilities is always one.

The instances of Scholl [1993] with more than 20 tasks are used to generate the instances for the presented problem. The instances are available for download at https://celso-sikora.com/publication-list or at https://www.bwl.uni-hamburg.de/or/team/celso-sikora.html. Although instances up to 297 tasks are generated and made available in the dataset, the largest instance solved by the proposed method has 53 tasks. The name of the instances considered for the result sections, the number of tasks, and the number of stations are summarized in Table 5.2. Out of a total of 272 instances generated for the dataset, 46 are solvable on a computer with an Intel i5 6500 processor at speed 3.20 GHz and 8 GB of RAM and an implementation of Visual Basic 2013. For each instance type (column Instance name), multiple instances using different numbers of stations and consequently different values of cycle time are generated. The generation of the options and processing times are identical within an instance type. In Table 5.2, the interval used to generate the number of stations is given. Furthermore, the last column of the table shows the number of possible products an assembly line could produce considering the combination of all available options.

Among the precedence diagram, options' processing time, their probabilities, and the number of stations, a cycle time is also defined for each instance. Both the number of stations and the cycle time are fixed so that the assembly line costs are based on the physical line length and the expected utility work. If either the number of workers or the cycle time is not fixed, their effect must also be accounted for in the cost function, generating a different class of problem. Every station must have an expected load less than or equal to the cycle time, otherwise, the station will surely be overloaded. Therefore, every solution for the problem must also be

Table 5.2 Summary of the instances of the dataset solved by the proposed method

Instance name	No. tasks	No. stations	No. of combinations
Buxey	29	7–13	$3.02 \cdot 10^7$
Gunther	35	6–9	$3.92 \cdot 10^{10}$
Hahn	53	3–10	$1.59 \cdot 10^{12}$
Heskia	28	3–8	$5.97 \cdot 10^6$
Lutz1	32	8–10	$1.45 \cdot 10^9$
Mitchell	21	3–8	62208
Roszieg	25	4–10	$2.52 \cdot 10^6$
Sawyer	30	7–11	$2.69 \cdot 10^7$

feasible in terms of cycle time considering the expected processing times of each task. In a preprocessing phase, all instances are solved as SALBP instances using the expected processing time of each task using the algorithm SALOME [Scholl and Klein, 1997]. The optimal solution or the best answer after 1,000 seconds is used in the calculation of the cycle time. The definition of the cycle time (CT) is given as

$$CT = Max \left(1.05 \cdot CT_{LB}; 1.02 \cdot CT_{UB}^{avg}\right)$$

in which CT_{LB} is a lower bound on the cycle time, given by

$$CT_{LB} = \frac{\sum_{t \in T} Dur_t^{avg}}{|S|}$$

and CT_{UB}^{avg} is the upper bound of the instance considering the expected processing times using SALOME. The selected cycle time for the instance under random sequences is slightly higher than the deterministic optimal. This way, the instance is defined within a reasonable production rate (at most 5% above the lower bound or 2% above a known upper bound, whichever is larger) but also allows flexibility to consider non-optimal deterministic answers. Without this extra flexibility, the deterministic solution may be the only feasible solution for the problem.

An instance also requires the specification of the line-length and utility-work cost coefficients: c_1 and c_2. As described in the problem definition, the speed of the conveyor is set to 1, so that distance and time are equivalent. For the instances, c_1 is considered to be 1 monetary unit per time unit. The length cost is due to the cost of acquisition and maintenance of the length of the conveyor system as well as the opportunity costs of the space used in the factory. The line-length cost is considered linear on the length of the stations and is calculated as the sum of all station costs. c_2 is equivalent to the price of a unit of time of utility work. The utility work cost calculation is based on the expected amount of utility work. The values of c_1 and c_2 are dependent on the physical layout, opportunity costs, wages, etc. For the dataset, the cost parameters are selected so that the resulting stations have a reasonable length. In preliminary tests values between 1 and 100 for c_2 generated. The optimal length depends on the selected cost coefficients since there is a trade-off between the station length and the expected utility work. Therefore, a set of representative values for the cost coefficients are selected for the dataset. In total, every instance of the dataset is solved with $c_1 = 1$ and $c_2 = 5, 10, 20,$ and 50. Smaller values of c_2 result in instances in which all stations lengths are equal to the lower bound. A value of $c_2 = 50$ already results in very small amounts of expected utility work, so that the selected range is considered representative for all possible scenarios. In total, the 46 base instances are solved with 4 different cost parameter combinations, resulting in 184 instances.

5.5.2 Results

The results containing the optimal value of the objective function, the line length (sum of all station lengths), the expected utility work, and the solution time are displayed in Tables 5.3, 5.5, and 5.6, respectively.

In Table 5.3, the total cost for each instance is displayed along with the optimal line length. For each instance type, the values of the number of stations and the corresponding cycle time are shown in each row of the table. As the model requires that the length of each station is at least the cycle time, the minimal length of the line is shown in column NS · CT. As $c_1 = 1$ for the dataset, the value of NS · CT is a lower bound for the total cost. The extra cost above this lower bound is divided between the cost of the expected utility work and any length unit added to any station. In Table 5.3 the results of each instance for all variations of the expected utility work cost parameter (c_2) are displayed side by side for each instance. The values are contained in columns $c_2 = 5, c_2 = 10, c_2 = 20,$ and $c_2 = 50$. As expected, the total cost increases with larger values for the cost parameter c_2. Furthermore, the more costly the utility work is, the longer the stations are planned. Note, however, that the effect of the different cost parameters differs from instance to instance. The setting $c_2 = 5$ results in almost no extra length for instances such as Buxey and Mitchell but yields differences of 1% to 4% in instances such as Heskia. For the most costly parameter ($c_2 = 50$), the optimal configuration exhibits up to 38% increase of the line length (Instance Heskia, 8 stations). As the increase of the line length is made in a discrete fashion (the length is assumed to be integer), the optimal length of a station is determined when the reduction of the expected utility work cost is less than the cost of one unit of length.

It is noteworthy, that the balancing solution for an instance is not necessarily the same for each cost parameter. Table 5.4 contains the information on the number of instances with different cost coefficients that have the same assignment solution. Out of the 46 instances, 21 instances have the same optimal balancing solution for every tested cost parameter. The optimal length varies according to the cost parameter. For the other 25 instances, not only the length of the line is different, but also the optimal task assignment itself. The optimal solutions of the instances are compared to each other and the number of identical assignments is displayed in Table 5.4. According to the values of Table 5.4, the number of identical optimal solutions is correlated with the difference of the cost parameters. For instance, 37 instances have the same optimal solution when solved for $c_2 = 10$ and $c_2 = 20$, while only 21 instances have identical assignments when considering the pair $c_2 = 5$ and $c_2 = 50$.

The relative values of the expected utility work are described in Table 5.5. The expected utility work values are presented in percentage of the cycle time. A value of 5%, for instance, means that it is expected that the line would require at least one utility worker 5% of the time during production. For the generated instances, the expected utility-work time varies from 0% (Buxey, 13 stations) to 22.43% (Heskia, 8 stations). The optimal value of the expected utility work strongly varies based on the cost parameter c_2. On average, the dataset requires 7.21% utility work time for the lowest cost setting ($c_2 = 5$) and only 1.01% for the highest value of c_2. In the problem formulation, the utility-work cost is based on the expected use of the utility worker. This is just a proxy for the utility-work cost since the number of utility workers is discrete. These workers, however, are shared between stations or even multiple assembly lines, so that only a proportion of their time can be accounted for as cost. Furthermore, the expected occupancy of a utility worker must be much less than 100%, since their necessity is not evenly spread through the production time; they need to move to the required workstation; there may be more than one station requiring utility work at the same time.

The trade-off between line length and expected utility work can be easily observed in the average values of Table 5.5. The extension of the line from an average of 3,809 time units ($c_2 = 5$) to 3,916 time units ($c_2 = 10$) reduces the expected utility work from an average of 7.21% to 3.91%. Further line extensions result in smaller effects on the expected utility-work average, indicating diminishing returns. Note that although the expected utility-work values seem small for the used cost

Table 5.3 Results of the optimal cost and length for the dataset

Instance	NS	NS-CT	$c_2 = 5$		$c_2 = 10$		$c_2 = 20$		$c_2 = 50$	
			Len	Cost	Len	Cost	Len	Cost	Len	Cost
Buxey	7	336	336	339.9	338	342.8	342	345.9	346	350.1
	8	336	336	342.3	341	347.3	346	351.4	350	357.1
	9	342	342	343.1	342	344.1	343	346.2	346	347.4
	10	340	340	343.9	342	347	346	348.3	348	351.4
	11	341	343	348.7	346	353.5	353	358.5	357	364.9
	12	348	348	351.9	352	354.2	353	356.1	356	358.8
	13	364	364	364	364	364	364	364	364	364
Gunther	6	462	475	526.1	504	566	546	607.5	601	665.6
	7	448	460	516.5	498	554.2	534	595.5	587	654.5
	8	448	458	518.4	499	560.5	539	602.8	595	665.7
	9	486	486	516	517	539.2	534	555.3	549	576.7
Hahn	3	13977	14030	14849.7	14664	15375.7	15069	15960.9	15832	16835.6
	4	14264	14366	14946.8	14739	15374.4	15184	15831.9	15705	16483.1
	5	13720	13852	14497.3	14202	14973.6	14764	15469.3	15349	16130.1
	6	14046	14640	15624.4	15092	16418.1	15962	17407.5	17111	18930.7
	7	14861	15369	16041.8	15603	16633.7	16238	17399.5	17111	18635.9
	8	14616	14692	15271.7	15092	15688.2	15444	16090.7	16074	16603
	9	14985	15257	15691.4	15560	16036.5	15960	16350.6	16270	16768.4
	10	15880	16306	16826.7	16670	17262.4	17158	17772.7	17646	18497.2

(continued)

Table 5.3 (continued)

Instance	NS	NS-CT	$c_2 = 5$		$c_2 = 10$		$c_2 = 20$		$c_2 = 50$	
			Len	Cost	Len	Cost	Len	Cost	Len	Cost
Heskia	3	981	999	1097.5	1058	1168.7	1112	1256.1	1221	1406.8
	4	984	997	1112.7	1070	1193.9	1134	1288.3	1255	1449.2
	5	985	996	1113.5	1070	1196.2	1137	1291.4	1255	1453.8
	6	984	999	1127.1	1077	1216.6	1142	1326.7	1279	1521.9
	7	987	1014	1141	1090	1235.3	1167	1346.3	1314	1541.8
	8	984	1027	1177.2	1125	1289.6	1195	1430.9	1362	1681.5
Lutz1	8	13944	14101	14884.5	14568	15482.9	15171	16172.2	16089	17138.9
	9	14355	14438	15010.2	14813	15413.8	15265	15835.2	15839	16377.1
	10	14640	14675	15124.3	14955	15487.2	15232	15876.6	15860	16527.6
Mitchell	3	108	108	112.4	111	115.8	115	119.4	119	123.8
	4	108	108	114.5	112	119.6	119	124.5	122	130
	5	110	110	115.7	114	120	119	124	123	128.9
	6	114	114	117.6	117	120.4	119	122.8	122	125.6
	7	112	112	118.8	118	122.4	120	125.5	126	130.2
	8	120	121	125.8	124	129.2	128	132.6	132	137.2
Roszieg	4	116	117	121.4	119	125.2	124	128.7	127	132.9
	5	115	115	122.7	121	127.5	126	132.2	132	138.7
	6	120	120	123.4	121	126.3	125	129	128	132.9
	7	126	127	129.4	128	131.3	131	133.4	133	135.9
	8	128	128	129.4	129	130.1	130	131	130	132.5
	9	117	119	126.6	122	132.5	129	138.8	136	149.3
	10	120	122	130.7	127	136.5	134	142.9	141	152.5
Sawyer	7	329	330	338.5	335	344.8	343	350.7	350	358.1
	8	328	329	341.5	339	349.4	344	357.1	356	368.7
	9	333	336	343.8	339	349.9	348	355.6	355	363.3
	10	330	330	343.2	340	350.2	344	358.5	356	368.8
	11	341	343	350.9	348	356.1	352	362.6	362	371.4
Average			3809	3986	3916	4119.7	4050	4269.7	4235	4491.7

Table 5.4 Number of instances with the same assignment for different combinations of the cost parameters. For example, 31 of the 46 instances with $c_2 = 5$ present the same balancing solution as for the instances with $c_2 = 10$

c_2	10	20	50
5	31	25	21
10		37	29
20			33

parameters, the instances contain from 3 to 11 stations. Longer lines (20 or more stations) would require more utility work for the same values of the cost parameter.

The cost parameter has a strong influence on the solution time of the instances, as described in Table 5.6. The proposed algorithm requires more time for larger values of c_2. The explanation of this clear trend lies in the average length of the stations. As the Markov chain matrices are built based on the extra length (difference of the station length and the cycle time), longer stations produce larger matrices. Therefore, larger cost parameters require the solution of larger linear systems, requiring more solution time on average. The difference can be extreme as in the instance Hahn with 6 stations, for which the optimal solution for $c_2 = 5$ is obtained in 2.3 seconds, while 614.7 seconds are required for $c_2 = 50$. On average, the solution times vary from 90.3 to 419.4 seconds. It is worthy to note, however, that several of the instances are very easy to solve, requiring much less than one second. Some instances of Hahn, which have long cycle times, however, require up to 8,195.4 seconds.

Table 5.5 Optimal length and percentage of utility work relative to the cycle time

Instance	NS	CT	$c_2 = 5$		$c_2 = 10$		$c_2 = 20$		$c_2 = 50$	
			Len	$\%UW$	Len	$\%UW$	Len	$\%UW$	Len	$\%UW$
Buxey	7	48	336	1.62	338	0.99	342	0.40	346	0.17
	8	42	336	2.99	341	1.50	346	0.64	350	0.34
	9	38	342	0.56	342	0.56	343	0.42	346	0.07
	10	34	340	2.27	342	1.46	346	0.34	348	0.20
	11	31	343	3.66	346	2.42	353	0.89	357	0.51
	12	29	348	2.68	352	0.75	353	0.53	356	0.19
	13	28	364	0	364	0	364	0	364	0
Gunther	6	77	475	13.28	504	8.05	546	3.99	601	1.68
	7	64	460	17.64	498	8.78	534	4.80	587	2.11
	8	56	458	21.57	499	10.98	539	5.70	595	2.53
	9	54	486	11.12	517	4.12	534	1.98	549	1.03
Hahn	3	4659	14030	3.52	14664	1.53	15069	0.96	15832	0.43
	4	3566	14366	3.26	14739	1.78	15184	0.91	15705	0.44
	5	2744	13852	4.70	14202	2.81	14764	1.29	15349	0.57
	6	2341	14640	8.41	15092	5.66	15962	3.09	17111	1.55
	7	2123	15369	6.34	15603	4.85	16238	2.74	17111	1.44
	8	1827	14692	6.35	15092	3.26	15444	1.77	16074	0.58
	9	1665	15257	5.22	15560	2.86	15960	1.17	16270	0.60
	10	1588	16306	6.56	16670	3.73	17158	1.94	17646	1.07

(continued)

Table 5.5 (continued)

Instance	NS	CT	$c_2 = 5$		$c_2 = 10$		$c_2 = 20$		$c_2 = 50$	
			Len	%UW	Len	%UW	Len	%UW	Len	%UW
Heskia	3	327	999	6.02	1058	3.39	1112	2.20	1221	1.14
	4	246	997	9.40	1070	5.04	1134	3.14	1255	1.58
	5	197	996	11.93	1070	6.41	1137	3.92	1255	2.02
	6	164	999	15.63	1077	8.51	1142	5.63	1279	2.96
	7	141	1014	18.01	1090	10.31	1167	6.36	1314	3.23
	8	123	1027	24.43	1125	13.38	1195	9.59	1362	5.19
Lutz1	8	1743	14101	8.99	14568	5.25	15171	2.87	16089	1.20
	9	1595	14438	7.17	14813	3.77	15265	1.79	15839	0.67
	10	1464	14675	6.14	14955	3.64	15232	2.20	15860	0.91
Mitchell	3	36	108	2.44	111	1.34	115	0.61	119	0.27
	4	27	108	4.85	112	2.82	119	1.02	122	0.59
	5	22	110	5.20	114	2.75	119	1.14	123	0.53
	6	19	114	3.83	117	1.77	119	1.00	122	0.37
	7	16	112	8.46	118	2.78	120	1.73	126	0.53
	8	15	121	6.42	124	3.45	128	1.55	132	0.69
Roszieg	4	29	117	3.05	119	2.13	124	0.82	127	0.41
	5	23	115	6.73	121	2.82	126	1.34	132	0.59
	6	20	120	3.44	121	2.67	125	1.00	128	0.49
	7	18	127	2.63	128	1.85	131	0.68	133	0.33
	8	16	128	1.80	129	0.67	130	0.31	130	0.31
	9	13	119	11.69	122	8.08	129	3.76	136	2.05
	10	12	122	14.43	127	7.89	134	3.71	141	1.92

(continued)

Table 5.5 (continued)

Instance	NS	CT	$c_2 = 5$		$c_2 = 10$		$c_2 = 20$		$c_2 = 50$	
			Len	%UW	Len	%UW	Len	%UW	Len	%UW
Sawyer	7	47	330	3.62	335	2.08	343	0.82	350	0.34
	8	41	329	6.08	339	2.53	344	1.59	356	0.62
	9	37	336	4.22	339	2.95	348	1.03	355	0.45
	10	33	330	8.01	340	3.09	344	2.20	356	0.77
	11	31	343	5.09	348	2.61	352	1.71	362	0.61
Average			3809	7.21	3916	3.91	4050	2.11	4235	1.01

Table 5.6 Comparison of the solution time of the dataset with respect to the optimal length

Instance	NS	CT	$c_2 = 5$		$c_2 = 10$		$c_2 = 20$		$c_2 = 50$	
			Len	Time	Len	Time	Len	Time	Len	Time
Buxey	7	48	336	0.3	338	0.4	342	0.4	346	0.3
	8	42	336	0.4	341	0.4	346	0.4	350	0.3
	9	38	342	0.4	342	0.4	343	0.6	346	0.4
	10	34	340	0.4	342	0.5	346	0.5	348	0.4
	11	31	343	0.3	346	0.3	353	0.3	357	0.3
	12	29	348	0.3	352	0.3	353	0.4	356	0.3
	13	28	364	0.4	364	0.4	364	0.4	364	0.4
Gunther	6	77	475	5.0	504	5.8	546	7.8	601	8.9
	7	64	460	1.9	498	2.4	534	3.2	587	5.3
	8	56	458	1.3	499	1.8	539	2.3	595	5.0
	9	54	486	3.1	517	3.4	534	3.5	549	4.8
Hahn	3	4659	14030	2.9	14664	66.0	15069	74.7	15832	155.5
	4	3566	14366	3025.2	14739	2754	15184	4884.6	15705	8195.4
	5	2744	13852	314.1	14202	974.6	14764	1453.5	15349	4320.3
	6	2341	14640	2.3	15092	9.7	15962	79.2	17111	614.7
	7	2123	15369	57.9	15603	111.6	16238	182.6	17111	701.0
	8	1827	14692	49.4	15092	74.5	15444	119.8	16074	350.5
	9	1665	15257	48.8	15560	102.8	15960	69.8	16270	114.3
	10	1588	16306	15.0	16670	22.8	17158	29.9	17646	70.7

(continued)

Table 5.6 (continued)

Instance	NS	CT	$c_2 = 5$		$c_2 = 10$		$c_2 = 20$		$c_2 = 50$	
			Len	Time	Len	Time	Len	Time	Len	Time
Heskia	3	327	999	298.1	1058	437.3	1112	798.1	1221	2128.9
	4	246	997	149.3	1070	214.4	1134	438.2	1255	1503.5
	5	197	996	98.5	1070	115.9	1137	192.2	1255	629.8
	6	164	999	39.8	1077	48.5	1142	63.9	1279	216.4
	7	141	1014	26.7	1090	34.2	1167	48.0	1314	142.7
	8	123	1027	3.4	1125	3.9	1195	4.4	1362	8.4
Lutz1	8	1743	14101	0.027	14568	0.7	15171	6.5	16089	34.7
	9	1595	14438	0.041	14813	0.4	15265	2.5	15839	14.5
	10	1464	14675	0.05	14955	1.1	15232	10.1	15860	57.3
Mitchell	3	36	108	0.010	111	0.012	115	0.014	119	0.017
	4	27	108	0.003	112	0.003	119	0.004	122	0.004
	5	22	110	0.002	114	0.002	119	0.002	123	0.002
	6	19	114	0.006	117	0.009	119	0.008	122	0.008
	7	16	112	0.005	118	0.006	120	0.006	126	0.007
	8	15	121	0.002	124	0.002	128	0.003	132	0.002
Roszieg	4	29	117	0.016	119	0.019	124	0.020	127	0.019
	5	23	115	0.008	121	0.009	126	0.021	132	0.010
	6	20	120	0.006	121	0.008	125	0.011	128	0.007
	7	18	127	0.024	128	0.025	131	0.060	133	0.026
	8	16	128	0.017	129	0.018	130	0.026	130	0.018
	9	13	119	0.002	122	0.002	129	0.002	136	0.002
	10	12	122	0.008	127	0.010	134	0.009	141	0.011
Sawyer	7	47	330	2.3	335	2.2	343	2.3	350	2.1
	8	41	329	1.8	339	1.9	344	2.0	356	2.1
	9	37	336	1.8	339	1.9	348	1.8	355	1.6
	10	33	330	1.5	340	1.4	344	1.4	356	1.4
	11	31	343	1.7	348	1.7	352	1.7	362	1.6
Average			3809	90.3	3916	108.6	4050	184.5	4235	419.4

Controlling Production Sequences Using Buffers

6

In the previous two chapters, both extreme assumptions on sequence control are dealt with for the balancing problem of mixed-model assembly lines. In Chapter 4, it is assumed that the sequencing of the models can be entirely controlled by the assembly line planner or operator. At the other end of the spectrum, no control over the sequencing is assumed in Chapter 5 so that the order of the produced pieces is completely random. In this chapter, an intermediate configuration is modeled, in which the planner has limited control over the sequence.

The boundary conditions and the level of control are described in the problem description section. For this problem, a simulation model, heuristic rules, and improvement procedures are introduced. This chapter presents the problem and initial results of the proposed improvement procedures.

6.1 Problem Description

The problem developed in this chapter consists of a constrained sequencing of the models produced in an automotive assembly line. As described in Chapter 2, the production sequence depends on different departments of a factory. The sales, marketing, production, and painting departments have different objective functions for a problem that is very interrelated between all these fields [Gagné et al. 2006]. The intersection of the different production sections is generally implemented with buffers. Such buffers provide some protection against small disruptions so that the whole plant is not affected. Furthermore, small sequence changes can be performed or restored in some buffer configurations.

Most of the literature on sequencing deals either with the painting or the assembly processes [Boysen et al. 2009c; Gagné et al. 2006]. Therefore, the buffer at the intersection of these two phases is selected as the focus of this chapter. Using the buffer capacity, a given sequence of products can be altered by a resequencing procedure.

The proposed problem consists of the optimization of the buffer operation. The space inside the buffers can be used to store products and alter the production sequence locally. This way, a sequencing more amenable for the final assembly can be achieved. Moreover, the input sequence of the buffer is considered to be stochastic. The sequencing decision of which product to send to the final assembly line is then performed under partial information. Only the state of the assembly line and the contents of the buffer are known so that the problem is online, i.e., it is solved repeatedly for each cycle.

6.1.1   Problem Classification and System Description

For this problem, it is assumed that the product sequence is independent of the assembly line objectives and can therefore be considered random with respect to the processing times. This assumption relies on the multi-objective nature of the sequencing problem, since several decision makers can influence the decision. Furthermore, the input of the buffer takes the products coming from the paint shop. According to Boysen et al. (2012), the paint shop is the least reliable part of the system, since small imperfections may require retouch or even a complete repainting of the product. Therefore, even planned sequences are randomly disturbed during the painting process.

Along with a random sequence of products, the buffer and the assembly line are considered to be given. The buffer has a given size, while the assignment of the tasks to the stations is already predefined. The optimization variable of this problem setting is the sequence of products that are sent to the assembly line. Based on the input sequence of the buffer and its capacity, the production sequence can be modified in a resequencing process.

The resequencing problem has been surveyed by Boysen et al. (2012), who propose a classification for this class of problem. The authors provide five classification criteria: the resequencing object, the resequencing objective, the solution approach, the planning horizon, and the resequencing trigger. The term resequencing object pertains to the question whether the products are resequenced physically or virtually. A virtual reschedule consists in reassigning the product to a different customer order, without a physical resequencing of the products [Boysen et al. 2012]. The physical resequencing configuration is further classified based on the buffer type. The second criterion is the resequencing objective, which is described as sequence restoration, mixed-model sequencing, car sequencing, level scheduling, and paint batching [Boysen et al. 2012]. For the restoration objective, the buffer is used to mitigate the errors in the process and to return the sequence to the previous original plan. The mixed-model sequencing takes into account the processing time and aims at the minimization of operational costs such as utility work, line stoppage, rework, etc. The car sequencing objective relates to sequencing rules in form H:N (at most H products of a given model can be present in a sequence of N vehicles). The level scheduling objective is used for the part storage and usage in the workstations. Finally, the paint batching optimizes the sequencing for the paint shop, for which set-up times or costs for different colors are assumed. The solution approach is the third criterion, which is divided into exact, heuristic, and simulation procedures. As the forth criterion, the planning horizon divides the literature into two categories: strategic and operational. The strategic problems cope with the sizing or positioning of the buffers, while the operational problems consider these variables as input parameters. The operational problems are further described based on the availability of information. The resequencing problems can be classified into two subgroups: static or deterministic, when all information is known; and dynamic or online, when the information is not completely available at the decision time. The last criterion is based on the resequencing trigger. Boysen et al. (2012) divide the trigger in reactive and proactive. The reactive problems aim at correcting sequences after disturbances and errors. The proactive procedures are used for the resequencing of products between production stages with different objective functions.

The boundary conditions of the proposed problem are now described following the classification on resequencing problems by Boysen et al. (2012). The products are physically resequenced using a buffer placed between the paint shop and the assembly line. The considered buffer type is AS/RS (Automated storage and retrieval system), depicted in Figure 6.1. This buffer provides the maximal resequencing flexibility for a given number of buffer positions. All vehicles are stored in slots from which they can be retrieved in any order.

The resequencing of the products is aimed at the optimization of the working effort in the stations of the assembly line. For that, the processing times are considered to minimize the amount of required utility work [Schumacher et al. 2018, Taube and Minner 2018]. Therefore, this problem's objective is described as mixed-model sequencing in the classification.

As the input sequencing of the buffer exiting the paint shop is considered random, a solution approach based on simulation is proposed. The resequencing problem is defined as an operational problem, since the buffer is considered given. The random

AS/RS

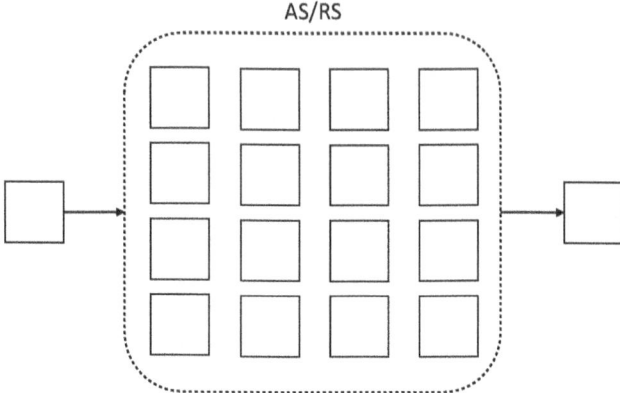

Figure 6.1 Automated storage and retrieval system (AS/RS) buffer

sequence implies that the products entering the buffer are not yet known, so that the decision is dynamic and online. The available computing time for such solution is very short, since a new model must be selected at each cycle time. For large manufacturers, the small cycle times may require a decision in less than a minute. As the information availability is incomplete and the solution time is very restricted, the proposed resequencing problem is said to be online [Boysen et al. 2012].

The final classification criterion does respect to the resequencing trigger. The buffer is used to shuffle the products to minimize the utility work for the assembly line. Therefore, the problem considers a proactive resequencing.

The problem definition also incorporates a due date for each product, so that the solution method does not hold heavily loaded models infinitely long in the buffer. The used due date is further explained in Subsection 6.1.2.

It is assumed that all the positions of the buffer can be used for the resequencing. This means that the buffer can operate all full capacity without having to assign positions to hedge for input or output delays.

6.1.2   Due Date Definition

According to the resequencing classification of Boysen et al. (2012) presented in Subsection 6.1.1, the minimization of operational costs in the final assembly is only one of the possible objective functions. As the automotive industry oft relies on Just-in-Time production, objectives considering the logistic of the assembled parts are also explored in the literature. For instance, Drexl et al. (2006) solve the sequencing problem considering both a car-sequencing problem and the level scheduling problem. In this chapter, only the minimization of the utility work is chosen as an objective function. However, the logistic of the required assembly parts is considered with a due date restriction. This due date avoids that products remain large periods in the buffer and is used as a proxy for the logistic restrictions.

The minimal due date restriction for a feasible operation of the buffer can be determined based on the First-in-First-out rule (FiFo). A new product enters the buffer in every cycle, while another product (which can also be the newly arrived product) is selected for production. Applying the FiFo rule for a buffer with B positions, a product will remain in the buffer $B - 1$ cycles (since the first model is sent at cycle 0). In order to allow any reordering flexibility for the resequencing problem, the products' due date must account for at least $B - 1$ cycles.

The due-date restriction is considered a hard constraint. That is, if the due date of a product has come, the product must be sent to production. This hard constraint simplifies the problem so that the objective function only relates to the utility work. If delays could be monetized, a combined objective function of production costs and delay costs could be used. Furthermore, a multi-objective problem definition is also possible.

6.1.3 Problem-State Definition, Transition Function, and Solution Policy

A state is defined based on all information available at the time of the decision. For a buffer of size B, each position of the buffer can be occupied by any of the combinations of options available. Each defined model has a given processing time in each station and a given due date, which must be met as a hard constraint. A state can be defined with the processing-time information and the amount of time until the due date for all positions. In the proposed dataset of Chapter 5, the instance named 'Hahn' contains $1.59 \cdot 10^{12}$ product combinations. Considering that each model also has a different due date, there are C^B load possibilities that can be encountered in the buffer, in which C is the number of possible products or option combinations.

The content of the buffer alone is not enough to fully describe a state. A decision on which model is sent to the line depends on whether the workers have enough time to process the tasks without needing utility work. Therefore, the state definition also contains the position of each worker in each station. As discussed in Chapter 5, the space between the cycle time equivalent and the station length contains all possibilities of the end position of a worker. For the instance 'Hahn' with 6 stations and utility-work cost (c_2) with a value of 50, the optimal length of the stations are 2341, 2529, 2893, 2493, 3197, and 3658. Given that the used cycle time is 2341, there are $1.67 \cdot 10^{13}$ combinations of worker positions in the stations.

Considering all factors, there are $C^B \cdot \prod_{s \in S}(Len_s - CT + 1)$ possible states which the system can assume, where S is the set of stations s, whose length is given by Len_s. Using the largest instance of Chapter 5, instance 'Hahn' with 6 stations and cost parameter $c_2 = 50$, the number of possible states surpasses 10^{135} considering a buffer with 10 positions. Such an instance is, however, small compared to real-world assembly-lines instances. An exact procedure for the problem should be able to select the best model to send to the assembly line in all these scenarios, given that the next model is random.

The transition from one state to a next state depends on the selected product that exits the buffer and a random product that enters the buffer. The selected product defines the position of the workers for the next state ($k + 1$) based on a state k in the form

$$Pos_s^{k+1} = \max \left(CT; \min \left(Pos_s^k + PT_p^k - CT; Len_s \right) \right). \tag{6.1}$$

The position of the worker in each station is bounded by the length of the station. Larger positions are avoided by requiring utility work. If a model requires a small processing time, idle time may occur, so that the minimal position is always the equivalent of the cycle time. In a physical assembly line, the product would require multiple cycle times to flow along the line, one station at a time. For state definition matters, the resulting position on all stations can be directly calculated as it only depends on the position and processing time within each station.

The due dates defined in Subsection 6.1.2 are related to a given time frame h. For the state definition, however, the present time is not important. A state with identical products, worker positions, and relative due dates is identical at any given time. A better definition for the due date (DD) of a product can be given in terms of multiples of cycle time left for its expiration. The transition function for the time left for the due date is given by $DD_p^{k+1} = DD_p^k - 1$, for each product p in cycle k.

A solution policy for the problem at hand has to provide a choice between the products in the buffer for any given possible state configuration. Based on the huge number of possible states, the lack of complete information for the decision, and the strong limit on the time for the decision, the problem is tackled here using heuristic rules or policies. These rules and policies are then evaluated by simulation.

6.2   Simulation Model

The simulation model used to evaluate the sequencing policies is illustrated in Figure 6.2. The simulator is divided into three parts: the model generation, the assembly line simulator, and the model selection.

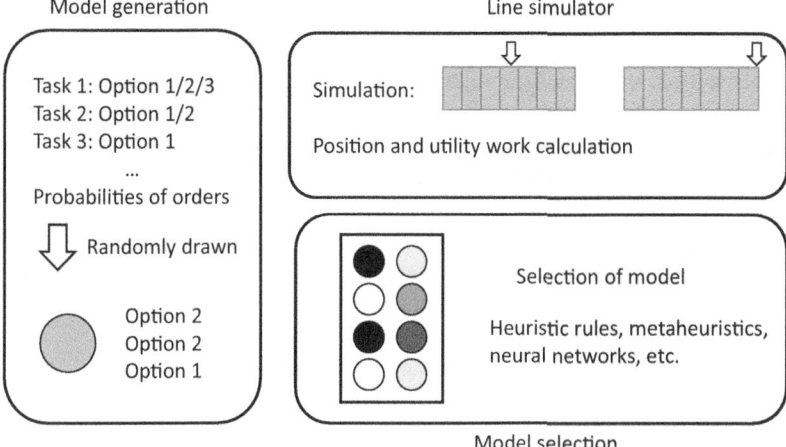

Figure 6.2 Structure of the buffer sequencing simulator

The model generator is responsible for the random input of products. The new product is drawn based on task options (as described in Chapter 5). A random number is drawn for each task, defining the selected option. The combination of all the options defines the product. Every time a product is selected to exit the buffer, a new entry product is generated.

The worker positions and the necessary utility work is calculated in the line simulator part. The position of the worker is calculated by Expression 6.1, while the utility work of a given product ($U W_{ps}^{k}$) is given by

$$U W_{ps}^{k} = \max \left(0,\ Pos_{s}^{k-1} + PT_{ps}^{k} - Len_{s} - CT \right).$$

The third element of the simulator is the decision policy. Here, it is decided which of the products within the buffer is sent to the assembly line. The data which are available for the decision are the processing times of each product in each station, the product due date, and the position of each worker. The selection must be made in real-time since a new vehicle is produced in every cycle time. For the selection of models, procedures such as heuristic rules, metaheuristics, and neural networks can be used. In this chapter, heuristic rules are presented in Section 6.3, while improvement approaches such as a local search and a lookahead search algorithm are presented in Section 6.4.

6.3   Heuristic Rules

In this section, some heuristic rules are proposed. The rules explore parts of the available information at the decision time and are used as a benchmark for other procedures. The heuristic rules are used as a product selection policy in the buffer and require little computation so that they can be employed in an online decision setting. For each rule, each product in the buffer receives a score, so that the product with the highest (or lowest) score is chosen and sent to the assembly line.

Since reversibility is desirable, the due dates of each product are modeled as hard constraints. Therefore, if a product in the buffer reaches its due date, it is selected independently from the preferred score given by the heuristic.

The heuristics that are used in the result section are:

- First-in-First-out
- Shortest total processing time
- Shortest total processing time on a subset of stations
- Smallest use of utility work
- Smallest amount of idle time.

The First-in-First-out (FiFo) rule sends the products exactly in the sequence they arrive. Therefore, the output sequence is also random and can be used as a reference value. The shortest total processing time rule selects the least loaded model in terms of the sum of the processing time in all stations. This selection corresponds to a greedy rule and is not expected to perform well.

The other three rules aim at better policies for buffer usage. The shortest total processing time on a subset of stations rule considers only the stations in which the worker is not at the left-most position. That is, the model with the shortest sum of processing times in the more critical stations is selected. The smallest use of utility work rule calculates the necessary utility work for all products in the buffer. The product that would cause the least amount of utility work is selected. Finally, the last explored heuristic policy selects the product with the minimal amount of idle time in the stations. In the case when more than one model ties for a given heuristic, another rule can be used as a tie-breaker. The due date can also be used for this purpose.

6.4 Improvement Approaches

As a solution to the online optimization problem, simple selection policies are suggested in Section 6.3. Now, two procedures are proposed: the first relies on characteristics of each model alone, while the second is based on partial sequences of the products in the buffer. The first policy calculates a score value for each product in the buffer, similar to the simple heuristics presented in Section 6.3. The score is defined with an expression using the available data at the decision moment: due date of each product, their processing times, and the position of the worker in each station. Instead of using a simple sum, the expression is defined containing weights and possible exponents and is optimized using a local search procedure. For this approach, the importance of the different factors in the expression used for the selection is modeled as the decision variables.

The second policy is a partial lookahead procedure. The lookahead ignores that new models will enter the buffer and calculates the best sequence for the production of the buffer content, up to a limit of the sequence length. As the decision must be performed in a short time (a product must be selected within each cycle time), only a partial enumeration of short sequences is considered.

6.4.1 Expression for the Product Selection Policy

The expression for the selection policy depends on the due date of the product and the respective worker positions. According to results presented later in Section 6.5 for the heuristic rules in Section 6.3, expressions based on the resulting utility work of a given selection are within the best rules tested. Generating an expression with the worker position does not produce good results, since the positions within the length of the station do not require utility work and should not be penalized.

The selected expression for the policies is given by

$$\text{Score}_p = \sum_{s=1}^{NS} K_s \cdot X_{ps}^{E_s},$$

in which a score (Score$_p$) is calculated for each product p. The modeled parameter in the rule is given by X. One possible rule is to consider X_{ps} to be the respective utility work for a given product p in station s. The score value is the sum of these factors, weighted based on the variables K for the linear coefficients, and E as the exponents.

The optimization procedure consists of choosing the best set of values K and E so that the selection policy produces the minimal amount of utility work. These variables are defined as continuous within an interval. For the coefficients, the interval $[-1; 1]$ is used, while the exponents are limited to $[-2; 2]$. The selection of the coefficients is made with a local-search algorithm.

6.4.2   Improvement Procedures

Local search
An approach for optimizing the score expression is based on testing neighboring solutions for a given set of values. Starting from a given or random solution, each parameter is tested independently. A value Δ is added and subtracted from each parameter, and the new rule is simulated. This way, the number of simulated answers per iteration is twice the number of variables. The best solution in the iteration is selected for the starting solution in the next iteration. The value of Δ is adjusted in each iteration, according to

$$\Delta = \frac{1}{1 + \bar{h}}$$

in which $\bar{h}$ refers to the number of iterations without improvement. The local search starts with a large value of Δ and refines the search throughout the iterations. The process is repeated until a given number of iterations without improvement.

Partial lookahead
The second improvement procedure is based on subsequences of the buffer's content. Instead of giving a score value for each product and selecting only one, the procedure considers combinations of multiple products.

Up to a depth b, all sub-sequences of the content of the buffer are tested. The sequencing with the lowest sum of utility work is selected, and its first model is sent to the final assembly line. In the next cycle, a new product arrives and the partial enumeration is started from scratch. That is, only the first model of the selected sequence is actually fixed. In order to meet the due dates, only sequences without delay are considered.

6.5 Tests and Results

For the tests, a dataset containing 45 instances ranging from 3 to 12 stations is used. An instance is composed of an assignment of tasks to stations and the length of each station. The tasks can exhibit multiple options, that is, alternative components to be built. The processing time and the relative probability of each option are also part of the instance definition. The combination of the options of the tasks with their given probabilities and processing times defines each of the possible product variations. For the dataset, the instances and solutions of Chapter 5 are used. The optimal solutions of 45 of the instances of Chapter 5 with parameter cost $c_2 = 10$ are selected for the simulation. The used line structure contains the optimal assignment of tasks for a random production sequence. By operating the buffer at the entry of the final assembly, the required amount of utility work can be expected to decrease. For all instances, a buffer with 10 positions is used. The selected value for the due date is twice the minimal required due date described in Subsection 6.1.2: $2 \cdot (B-1)$. This due date is equal to every model entering the buffer.

The result section contains details on the simulation length and the number of repetitions. Furthermore, the solution quality of the heuristic rules is compared with the proposed improvement procedures. All instances and results of this chapter are available again at https://www.bwl.uni-hamburg.de/or/team/celso-sikora.html or https://celso-sikora.com/publication-list.

6.5.1 Simulation Length and Number of Repetitions

The aim of this chapter lays on the evaluation of the heuristic rules and the improvement procedures in terms of required utility work. The tested algorithms are solution proposals for which the solution time is not as important as to show their potential. Therefore, the simulation length and the number of repetitions are set to a value high enough, so that the confidence intervals are small and comparisons are straightforward.

For the validation of the simulation and selection of the simulation length, the simulated assembly line is compared with the results of Chapter 5. In Chapter 5, the expected value of the utility work for a random sequence is calculated with the help of Markov chains. Therefore, the simulated results using a FiFo rule can be expected to be similar to the exact results provided in Chapter 5.

In Table 6.1 simulations of the dataset with different lengths (number of simulated products) are compared. Using 500 repetitions, the simulations are run for a range of 500 to 100,000 products. Column 'UW' contains the expected utility work obtained exactly using Markov chains. The utility work is expressed in 'Utility work time units per cycle'. The numerical columns contain the relative error (in percent) of the simulation compared to the expected value. The last two lines of the table contain the maximal and average values. As expected, the difference between the simulated values and the exact value diminishes with increasing simulation length. In general, the simulation is accurate even with very small sequences: the average error using a length of only 500 products is on average 0.373%. For fine comparisons, however, longer simulation runs may be required.

The variance of the simulated results is also important for the comparison of the two policies. Based on the standard deviation of the simulation runs and the number of repetitions, a confidence interval can be calculated for the average utility work value. The half-width of the 95%-confidence intervals based on a normal distribution for all instances with 500 repetitions is displayed in Table 6.2. The half-widths are displayed relative to the average value in percentage. An entry of 1.0%, for instance, represents a confidence interval of $[0.99\mu; 1.01\mu]$, in which μ is the average value. As expected, the confidence intervals are smaller with an increase in the simulation length.

Based on the results of Tables 6.1 and 6.2, the selected simulation length for all results shown in the Table 6.3 is of 50,000 products. This selection produces a very low error from the expected value calculated exactly with the Markov chain (Chapter 5) and has very narrow confidence intervals. This length is selected with the expectation that the different methods show large quality differences. Furthermore, it should be expected that the confidence interval for rules other than FiFo to vary, since selection rules bias the produced sequence. For the design of efficient solution algorithms, the simulation length may be reduced or selected dynamically. Since the local search requires a large number of simulations, the simulation length is reduced to 100 iterations of 10,000 products. The solutions found by these methods are then re-simulated using 50,000 products for a better comparison.

6.5.2 Implementation Details

The simulator, the heuristics, and the improvement procedures are implemented in C# and are run on an Intel i7-8700K at 3.7 GHz using up to 6 cores and 32 GB RAM. The simulations are run in parallel and require between 1.5 and 6 seconds for 500 repetitions of 50,000 products using the FiFo rule. The local search stopping criteria is set for 20 iterations without any improvement or a total of 200 iterations. For all tests, a warm-up period of 500 products is used.

Table 6.1 Relative error of the utility work value for 500 repetitions of simulation lengths of 500 to 100,000

Instance	NS	CT	UW (exact)	Relative abs. error (in %) - Simulation length (thous.)							
				0.5	1	2.5	5	10	25	50	100
Buxey	7	48	0.477	0.237	0.131	0.544	0.059	0.018	0.035	0.033	0.018
	8	42	0.629	0.327	0.032	0.039	0.051	0.021	0.113	0.013	0.071
	9	38	0.214	0.964	0.394	0.912	0.009	0.113	0.016	0.109	0.087
	10	34	0.496	0.179	0.899	0.238	0.175	0.038	0.017	0.062	0.062
	11	31	0.751	0.299	0.255	0.120	0.067	0.017	0.027	0.052	0.025
	12	29	0.217	0.700	0.877	0.317	0.183	0.516	0.126	0.080	0.002
Gunther	6	77	6.202	0.006	0.284	0.168	0.261	0.078	0.001	0.045	0.029
	7	64	5.621	0.238	0.380	0.203	0.009	0.031	0.038	0.063	0.001
	8	56	6.148	0.557	0.124	0.009	0.072	0.074	0.034	0.008	0.002
	9	54	2.223	0.404	0.477	0.032	0.021	0.199	0.077	0.055	0.022
Hahn	3	4659	71.167	0.380	0.682	0.016	0.080	0.213	0.118	0.003	0.012
	4	3566	63.537	0.081	0.561	0.061	0.138	0.073	0.045	0.052	0.018
	5	2744	77.16	0.688	0.005	0.146	0.018	0.088	0.055	0.012	0.008
	6	2341	132.612	0.001	0.027	0.006	0.065	0.100	0.013	0.012	0.014
	7	2123	103.069	0.387	0.479	0.193	0.238	0.065	0.055	0.012	0.029
	8	1827	59.619	0.289	0.052	0.129	0.087	0.016	0.102	0.031	0.022
	9	1665	47.647	0.192	0.439	0.11	0.153	0.078	0.063	0.038	0.043
	10	1588	59.243	0.514	0.001	0.178	0.012	0.081	0.079	0.049	0.002
Heskia	3	327	11.075	0.205	0.365	0.113	0.115	0.036	0.095	0.081	0.063
	4	246	12.389	0.291	0.048	0.005	0.198	0.064	0.022	0.021	0.051
	5	197	12.621	0.021	0.083	0.044	0.184	0.053	0.001	0.045	0.026
	6	164	13.96	0.024	0.003	0.004	0.031	0.068	0.049	0.006	0.009
	7	141	14.531	0.281	0.172	0.124	0.055	0.001	0.01	0.049	0.018
	8	123	16.46	0.570	0.158	0.143	0.036	0.085	0.033	0.015	0.03

(continued)

Table 6.1 (continued)

Instance	NS	CT	UW (exact)	Relative abs. error (in %) - Simulation length (thous.)							
				0.5	1	2.5	5	10	25	50	100
Lutz1	8	1743	91.491	0.204	0.067	0.032	0.080	0.063	0.029	0.073	0.019
	9	1595	60.085	0.561	0.412	0.101	0.082	0.062	0.024	0.033	0.004
	10	1464	53.219	0.143	0.214	0.029	0.096	0.051	0.074	0.041	0.006
Mitchell	3	36	0.482	0.578	0.085	0.080	0.091	0.101	0.043	0.057	0.058
	4	27	0.762	0.352	0.196	0.008	0.037	0.091	0.074	0.008	0.003
	5	22	0.604	0.161	0.076	0.762	0.247	0.154	0.079	0.001	0.004
	6	19	0.336	0.130	0.731	0.043	0.268	0.015	0.106	0.012	0.051
	7	16	0.444	0.417	0.082	0.033	0.012	0.031	0.026	0.069	0.008
	8	15	0.518	0.161	0.028	0.101	0.184	0.126	0.066	0.051	0.025
Roszieg	4	29	0.617	0.185	0.118	0.131	0.092	0.257	0.026	0.007	0.034
	5	23	0.648	0.846	0.012	0.185	0.100	0.020	0.063	0.017	0.067
	6	20	0.533	0.146	0.258	0.208	0.050	0.087	0.050	0.073	0.004
	7	18	0.333	1.114	0.524	0.644	0.264	0.229	0.006	0.005	0.034
	8	16	0.107	1.761	0.303	0.255	0.150	0.192	0.110	0.014	0.027
	9	13	1.050	0.308	0.089	0.199	0.057	0.072	0.031	0.022	0.012
	10	12	0.947	0.078	0.690	0.082	0.031	0.039	0.026	0.024	0.007
Sawyer	7	47	0.977	0.313	0.035	0.099	0.238	0.065	0.004	0.034	0.027
	8	41	1.037	0.593	0.321	0.316	0.007	0.037	0.031	0.038	0.007
	9	37	1.092	0.259	0.102	0.061	0.046	0.088	0.046	0.049	0.027
	10	33	1.018	0.231	0.038	0.077	0.158	0.062	0.034	0.026	0.012
	11	31	0.809	0.411	0.136	0.002	0.174	0.004	0.013	0.023	0.036
Max				1.761	0.899	0.912	0.268	0.516	0.126	0.109	0.087
Average				0.373	0.254	0.162	0.106	0.088	0.048	0.036	0.025

Table 6.2 95%-half-confidence interval size of the of each simulation in percentage of the average value for different simulation lengths

Instance	NS	CT	Relative size of a half-confidence interval (%)							
			0.5	1	2.5	5	10	25	50	100
Buxey	7	48	0.987	0.700	0.464	0.315	0.230	0.152	0.108	0.075
	8	42	0.956	0.753	0.478	0.324	0.211	0.135	0.099	0.074
	9	38	1.375	0.895	0.560	0.421	0.292	0.179	0.129	0.090
	10	34	0.779	0.606	0.371	0.276	0.183	0.119	0.083	0.060
	11	31	0.730	0.517	0.322	0.222	0.160	0.102	0.070	0.049
	12	29	1.390	0.989	0.593	0.446	0.315	0.205	0.146	0.100
Gunther	6	77	0.749	0.531	0.315	0.227	0.172	0.106	0.077	0.053
	7	64	0.839	0.537	0.359	0.238	0.172	0.112	0.080	0.057
	8	56	0.741	0.541	0.313	0.227	0.164	0.098	0.070	0.051
	9	54	0.784	0.578	0.369	0.255	0.186	0.116	0.080	0.058
Hahn	3	4659	1.073	0.809	0.487	0.373	0.227	0.156	0.112	0.074
	4	3566	0.777	0.541	0.350	0.260	0.175	0.111	0.082	0.054
	5	2744	0.708	0.490	0.304	0.221	0.155	0.098	0.071	0.050
	6	2341	0.621	0.424	0.274	0.193	0.139	0.087	0.059	0.042
	7	2123	0.787	0.564	0.336	0.268	0.177	0.116	0.075	0.053
	8	1827	0.722	0.514	0.322	0.222	0.162	0.101	0.073	0.052
	9	1665	0.814	0.596	0.363	0.259	0.187	0.119	0.079	0.058
	10	1588	0.997	0.687	0.415	0.318	0.216	0.139	0.096	0.065
Heskia	3	327	0.861	0.596	0.381	0.265	0.202	0.135	0.093	0.065
	4	246	0.745	0.522	0.336	0.236	0.169	0.103	0.080	0.055
	5	197	0.781	0.501	0.336	0.228	0.165	0.102	0.073	0.052
	6	164	0.720	0.498	0.330	0.218	0.158	0.100	0.073	0.052
	7	141	0.556	0.413	0.267	0.184	0.126	0.085	0.061	0.041
	8	123	0.589	0.417	0.250	0.183	0.136	0.085	0.063	0.043

(continued)

Table 6.2 (continued)

Instance	NS	CT	Relative size of a half-confidence interval (%)							
			0.5	1	2.5	5	10	25	50	100
Lutz1	8	1743	0.652	0.481	0.285	0.216	0.146	0.090	0.065	0.046
	9	1595	0.748	0.519	0.331	0.246	0.169	0.106	0.075	0.051
	10	1464	0.674	0.485	0.306	0.215	0.155	0.101	0.069	0.049
Mitchell	3	36	1.036	0.700	0.469	0.331	0.233	0.149	0.102	0.074
	4	27	0.774	0.553	0.350	0.251	0.174	0.113	0.078	0.056
	5	22	0.887	0.658	0.409	0.280	0.196	0.121	0.090	0.064
	6	19	1.007	0.715	0.432	0.312	0.219	0.140	0.102	0.068
	7	16	0.838	0.593	0.370	0.260	0.188	0.119	0.087	0.059
	8	15	0.879	0.624	0.387	0.270	0.197	0.121	0.085	0.061
Roszieg	4	29	0.838	0.605	0.359	0.259	0.186	0.115	0.079	0.054
	5	23	0.997	0.692	0.440	0.332	0.220	0.140	0.096	0.070
	6	20	0.735	0.532	0.337	0.233	0.170	0.101	0.075	0.051
	7	18	1.166	0.815	0.496	0.352	0.253	0.165	0.107	0.076
	8	16	1.575	1.120	0.696	0.474	0.337	0.204	0.152	0.110
	9	13	0.614	0.429	0.273	0.190	0.135	0.081	0.059	0.041
	10	12	0.640	0.437	0.278	0.193	0.137	0.089	0.062	0.044
Sawyer	7	47	0.815	0.572	0.363	0.253	0.180	0.112	0.081	0.056
	8	41	0.811	0.636	0.377	0.273	0.188	0.119	0.085	0.059
	9	37	0.791	0.565	0.349	0.257	0.176	0.111	0.079	0.054
	10	33	0.814	0.585	0.363	0.261	0.179	0.112	0.081	0.055
	11	31	0.800	0.574	0.350	0.248	0.172	0.109	0.075	0.055
Max			1.575	1.120	0.696	0.474	0.337	0.205	0.152	0.110
Average			0.848	0.602	0.376	0.269	0.189	0.120	0.085	0.059

6.5.3 Quality of the Heuristic Rules

The 95%-confidence intervals of simulations with 500 repetitions for each heuristic policy are given in Table 6.3. The difference in performance is very uniform for the different rules. Although all rules have roughly the same value for a couple of instances, the instances with different performances show a clear ranking among the heuristics. The first-in-first-out ('FiFo') rule is used as a comparison since it is, in theory, equivalent to a random sequence. Ordering products by ascending sums of processing times ('shortest PT') slightly improves the average utility work values (39 out of the 45 confidence intervals do not overlap with the FiFo rule). This policy, however, is not very promising, since the more loaded models accumulate in the buffer and must leave at their due date at the latest. An improvement is observed when the sum of the processing times is only performed at stations with a worker position larger than zero ('St. sm. PT', or Station-specific smallest sum of the processing time). Using this rule, the buffer would send products with short durations on the most critical stations, while the processing times in the other stations are irrelevant. The second better performing rule, on average, is the minimization of the idle time ('min IT'). Although this rule has a lower average utility work for the whole dataset, it is not better than the 'St. sm. PT' rule for several instances (for instance, 'Hahn' with 8 stations). The reasoning behind this rule is to avoid idle times at the stations so that the utilization of the workers is high. Finally, the best performing rule of the set is a greedy rule minimizing the utility work of each assignment itself ('min UW'). The selected model is the one that causes no utility work or the minimal amount among the options. Besides two instances ('Buxey' with 9 stations and 'Roszieg' with 6 stations) in which the difference is within the confidence intervals, the minimal utility work policy performs better in all other 43 instances. On average, the required utility work reduced from 20.785 for a random sequence to 12.380 when using this minimal-utility-work rule. These results show the cost-reduction potential in assembly lines when a buffer is used to resequence the production input. The improvement varies greatly between the instances: there is no improvement for 'Buxey' with 9 stations, while a difference of 87.7% (based on the average values) is observed in 'Hahn' with 10 stations.

As the Minimal-Utility-Work Rule ('min UW') performs better than the other tested ones, it is used as a base for the improvement procedures.

Table 6.3 Confidence interval of the heuristic policies for the average utility work per cycle for each instance of the dataset

Instance	NS	CT	Utility work - Rule				
			FiFo	Shortest PT	St. sm. PT	min IT	min UW
Buxey	7	48	[0.477, 0.479]	[0.475, 0.477]	[0.440, 0.442]	[0.464, 0.466]	[0.419, 0.421]
	8	42	[0.628, 0.631]	[0.623, 0.625]	[0.534, 0.537]	[0.582, 0.585]	[0.492, 0.494]
	9	38	[0.213, 0.214]	[0.213, 0.214]	[0.213, 0.214]	[0.213, 0.215]	[0.213, 0.215]
	10	34	[0.495, 0.497]	[0.495, 0.497]	[0.470, 0.471]	[0.488, 0.489]	[0.462, 0.464]
	11	31	[0.750, 0.752]	[0.742, 0.744]	[0.624, 0.626]	[0.701, 0.703]	[0.478, 0.480]
	12	29	[0.217, 0.218]	[0.214, 0.216]	[0.163, 0.164]	[0.185, 0.186]	[0.086, 0.087]
Gunther	6	77	[6.20, 6.22]	[6.08, 6.09]	[5.55, 5.56]	[5.52, 5.54]	[4.85, 4.87]
	7	64	[5.61, 5.63]	[5.48, 5.50]	[4.74, 4.75]	[4.63, 4.64]	[3.98, 4.00]
	8	56	[6.14, 6.15]	[6.01, 6.03]	[5.28, 5.30]	[5.18, 5.20]	[4.70, 4.72]
	9	54	[2.22, 2.23]	[2.18, 2.19]	[1.74, 1.75]	[1.99, 2.00]	[1.32, 1.33]
Hahn	3	4659	[70.86, 71.22]	[67.10, 67.40]	[65.29, 65.63]	[53.96, 54.25]	[44.77, 45.04]
	4	3566	[63.48, 63.68]	[62.79, 62.99]	[57.65, 57.87]	[59.20, 59.41]	[49.02, 49.20]
	5	2744	[77.07, 77.31]	[76.14, 76.37]	[71.51, 71.72]	[73.02, 73.22]	[60.85, 61.04]
	6	2341	[132.4, 132.8]	[128.5, 128.8]	[126.1, 126.4]	[91.36, 91.69]	[64.44, 64.71]
	7	2123	[103.0, 103.3]	[99.08, 99.40]	[89.77, 90.10]	[70.42, 70.73]	[26.01, 26.27]
	8	1827	[59.62, 59.8]	[58.9, 59.09]	[50.65, 50.8]	[56.00, 56.17]	[43.95, 44.11]
	9	1665	[47.53, 47.7]	[45.33, 45.49]	[38.42, 38.57]	[36.18, 36.31]	[16.95, 17.05]
	10	1588	[59.18, 59.44]	[55.67, 55.89]	[42.15, 42.40]	[34.54, 34.73]	[7.26, 7.34]

(continued)

Table 6.3 (continued)

Instance	NS	CT	Utility work - Rule				
			FiFo	Shortest PT	St. sm. PT	min IT	min UW
Heskia	3	327	[11.05, 11.09]	[10.51, 10.55]	[9.94, 9.99]	[9.08, 9.13]	[6.20, 6.24]
	4	246	[12.40, 12.43]	[11.92, 11.97]	[11.39, 11.44]	[10.83, 10.87]	[7.67, 7.71]
	5	197	[12.63, 12.66]	[12.17, 12.21]	[11.97, 12.01]	[11.15, 11.2]	[8.37, 8.40]
	6	164	[13.93, 13.97]	[13.56, 13.60]	[12.65, 12.69]	[11.79, 11.83]	[8.92, 8.96]
	7	141	[14.50, 14.54]	[14.43, 14.46]	[13.71, 13.75]	[13.37, 13.41]	[10.24, 10.28]
	8	123	[16.45, 16.49]	[16.29, 16.33]	[15.04, 15.09]	[11.97, 12.01]	[10.20, 10.24]
Lutz1	8	1743	[91.32, 91.59]	[90.08, 90.32]	[88.54, 88.76]	[81.52, 81.73]	[71.37, 71.60]
	9	1595	[59.98, 60.18]	[58.92, 59.09]	[55.5, 55.67]	[53.81, 53.99]	[47.51, 47.68]
	10	1464	[53.16, 53.3]	[53.06, 53.22]	[47.49, 47.62]	[51.92, 52.07]	[46.80, 46.95]
Mitchell	3	36	[0.482, 0.484]	[0.479, 0.481]	[0.447, 0.449]	[0.461, 0.463]	[0.414, 0.416]
	4	27	[0.761, 0.763]	[0.756, 0.759]	[0.696, 0.698]	[0.727, 0.729]	[0.664, 0.667]
	5	22	[0.603, 0.606]	[0.599, 0.601]	[0.558, 0.560]	[0.570, 0.573]	[0.527, 0.529]
	6	19	[0.335, 0.337]	[0.334, 0.335]	[0.31, 0.312]	[0.319, 0.321]	[0.271, 0.272]
	7	16	[0.443, 0.445]	[0.438, 0.440]	[0.344, 0.345]	[0.384, 0.385]	[0.268, 0.269]
	8	15	[0.517, 0.519]	[0.515, 0.517]	[0.401, 0.403]	[0.473, 0.474]	[0.229, 0.230]
Roszieg	4	29	[0.616, 0.618]	[0.611, 0.613]	[0.571, 0.574]	[0.582, 0.584]	[0.537, 0.539]
	5	23	[0.646, 0.649]	[0.628, 0.630]	[0.538, 0.541]	[0.526, 0.528]	[0.447, 0.449]
	6	20	[0.533, 0.534]	[0.532, 0.534]	[0.515, 0.517]	[0.526, 0.528]	[0.514, 0.516]
	7	18	[0.333, 0.334]	[0.330, 0.332]	[0.239, 0.241]	[0.306, 0.308]	[0.117, 0.118]
	8	16	[0.107, 0.107]	[0.105, 0.106]	[0.072, 0.073]	[0.097, 0.098]	[0.024, 0.025]
	9	13	[1.05, 1.05]	[1.04, 1.04]	[0.907, 0.910]	[0.962, 0.965]	[0.675, 0.678]
	10	12	[0.946, 0.949]	[0.931, 0.933]	[0.744, 0.746]	[0.821, 0.824]	[0.538, 0.540]
Sawyer	7	47	[0.975, 0.978]	[0.967, 0.970]	[0.862, 0.865]	[0.931, 0.934]	[0.798, 0.801]
	8	41	[1.03, 1.04]	[1.02, 1.02]	[0.865, 0.868]	[0.936, 0.940]	[0.727, 0.731]
	9	37	[1.09, 1.09]	[1.08, 1.09]	[0.968, 0.972]	[1.02, 1.03]	[0.764, 0.767]
	10	33	[1.02, 1.02]	[1.01, 1.01]	[0.774, 0.776]	[0.931, 0.935]	[0.541, 0.544]
	11	31	[0.809, 0.811]	[0.803, 0.806]	[0.598, 0.601]	[0.749, 0.752]	[0.318, 0.320]
Average			20.785	20.234	18.652	16.951	12.380

6.5.4   Improving Expression-Based Rules Using Local Search

In this section, local search is used to improve the results of expression-based policies. The idea behind the procedure is to change one of the expression's parameters
at a time and simulate the behavior of the buffer. Whenever an improvement is
found, the procedure repeats with the new incumbent solution. The search width Δ
diminishes with the iterations until a number of iterations without improvement is
executed.

Two experiments are proposed for the comparison. In the first, the local search
algorithm starts with the 'min UW'-rule values. For the second experiment, random start values are used. The expectation to achieve better expression rules is that
not all stations contribute equally to the total amount of utility work. The expected
occupation and the variation of the processing time in each station are not identical. Therefore, different weights in the expressions may provide better results than
choosing the product by a simple sum of utility work.

Table 6.4 contains the results of both experiments in comparison to the 'min
UW' rule. The column 'short run' refers to the average amount of utility work
obtained by the local search using the reduced simulation length of 100 repetitions
of 10,000 products. The final solutions are re-simulated with 500 repetitions of
50,000 products in the column 'long run', for which the 95% confidence intervals
are given.

The local search with the short runs yields improvements of more than 1% for 42
out of the 45 instances. Evaluating the same solutions with longer simulations shows
a different result: an improvement of 1% is only present in 28 out of the 45 instances.
In no case is the average utility work better in the long simulation. This illustrates
a difficulty of optimization using simulation: what appears to be the best solution
throughout the algorithm is very probably only an optimistic solution. Since a lot of
combinations are tested and only the best is selected, it is natural that simulations
with outliers stand out. In the literature robust optimization with simulation, Beyer
and Sendhoff (2007) discuss the use of reevaluating good solutions to reduce this
risk. In general, the local search starting with the 'min UW' expression can be
slightly improved on average with the proposed procedure. In comparison with the
results of the simulation runs of 500 repetitions, the average utility work for the
dataset reduces from 12.38 to 11.97.

The explanation of an improvement is illustrated with the instance 'Hahn' with
6 stations, which presents the largest improvement applying the local search. The
resulting expression for the instance is given by

$$\text{Score}_p = 0.952 \cdot UW_{p1}^1 + 0 \cdot UW_{p2} + 1 \cdot UW_{p3}^1 + 1 \cdot UW_{p4}^{0.5} + 1 \cdot UW_{p5}^1 + 1 \cdot UW_{p6}^{0.947},$$

Table 6.4 Comparison of the minimal utility work rule with the expression rules found by local search

Instance	NS	CT	min UW	Local search from min UW		Local search random start	
				short run	long run	short run	long run
Buxey	7	48	[0.419, 0.421]	0.411	[0.415, 0.417]	0.616	[0.618, 0.620]
	8	42	[0.492, 0.494]	0.489	[0.493, 0.495]	0.883	[0.887, 0.889]
	9	38	[0.213, 0.215]	0.211	[0.213, 0.214]	0.211	[0.214, 0.214]
	10	34	[0.462, 0.464]	0.459	[0.462, 0.463]	0.460	[0.464, 0.465]
	11	31	[0.478, 0.480]	0.466	[0.469, 0.471]	1.00	[1.00, 1.01]
	12	29	[0.086, 0.087]	0.085	[0.086, 0.087]	0.373	[0.377, 0.378]
Gunther	6	77	[4.85, 4.87]	4.79	[4.81, 4.83]	5.52	[5.55, 5.57]
	7	64	[3.98, 4.00]	3.94	[3.97, 3.98]	5.39	[5.41, 5.43]
	8	56	[4.70, 4.72]	4.61	[4.64, 4.65]	4.76	[4.79, 4.81]
	9	54	[1.33, 1.33]	1.27	[1.28, 1.28]	2.78	[2.79, 2.80]
Hahn	3	4659	[44.78, 45.04]	44.29	[44.76, 44.92]	44.46	[44.75, 44.91]
	4	3566	[49.02, 49.20]	47.44	[48.71, 48.85]	66.34	[69.84, 69.99]
	5	2744	[60.85, 61.04]	58.83	[59.13, 59.30]	77.56	[78.22, 78.42]
	6	2341	[64.44, 64.71]	57.30	[57.78, 58.01]	82.63	[83.41, 83.66]
	7	2123	[26.01, 26.27]	25.57	[26.06, 26.23]	78.23	[78.67, 78.91]
	8	1827	[43.95, 44.11]	41.66	[41.93, 42.06]	59.71	[59.90, 60.06]
	9	1665	[16.95, 17.05]	13.69	[16.47, 16.54]	15.75	[15.88, 15.94]
	10	1588	[7.26, 7.34]	6.96	[7.29, 7.36]	136.94	[137.32, 137.5]

(continued)

Table 6.4 (continued)

Instance	NS	CT	min UW	Local search from min UW		Local search random start	
				short run	long run	short run	long run
Heskia	3	327	[6.20, 6.24]	5.83	[6.16, 6.18]	6.11	[6.18, 6.20]
	4	246	[7.67, 7.71]	7.37	[7.42, 7.44]	8.23	[8.29, 8.32]
	5	197	[8.37, 8.40]	7.92	[7.98, 8.01]	8.68	[8.73, 8.76]
	6	164	[8.92, 8.96]	8.62	[8.68, 8.71]	9.49	[9.58, 9.61]
	7	141	[10.24, 10.28]	9.80	[9.85, 9.87]	12.32	[12.38, 12.41]
	8	123	[10.20, 10.24]	9.97	[10.03, 10.06]	11.27	[11.33, 11.37]
Lutz1	8	1743	[71.37, 71.60]	68.70	[68.99, 69.19]	90.15	[90.58, 90.81]
	9	1595	[47.51, 47.68]	45.59	[45.82, 45.96]	54.89	[55.17, 55.33]
	10	1464	[46.81, 46.95]	45.55	[45.79, 45.96]	55.50	[55.71, 55.87]
Mitchell	3	36	[0.414, 0.416]	0.408	[0.411, 0.412]	0.492	[0.494, 0.496]
	4	27	[0.664, 0.667]	0.651	[0.655, 0.657]	0.718	[0.722, 0.724]
	5	22	[0.527, 0.529]	0.514	[0.518, 0.519]	0.615	[0.619, 0.621]
	6	19	[0.271, 0.272]	0.261	[0.264, 0.264]	0.264	[0.266, 0.267]
	7	16	[0.268, 0.269]	0.252	[0.254, 0.255]	0.709	[0.711, 0.713]
	8	15	[0.229, 0.230]	0.200	[0.225, 0.226]	0.214	[0.216, 0.217]
Roszieg	4	29	[0.537, 0.539]	0.526	[0.528, 0.529]	0.772	[0.779, 0.78]
	5	23	[0.447, 0.449]	0.430	[0.434, 0.436]	0.855	[0.861, 0.863]
	6	20	[0.514, 0.516]	0.512	[0.515, 0.517]	0.511	[0.514, 0.515]
	7	18	[0.117, 0.118]	0.111	[0.113, 0.114]	0.564	[0.566, 0.567]
	8	16	[0.024, 0.025]	0.023	[0.023, 0.024]	0.023	[0.024, 0.025]
	9	13	[0.675, 0.678]	0.609	[0.611, 0.613]	0.850	[0.855, 0.857]
	10	12	[0.538, 0.540]	0.528	[0.537, 0.539]	0.725	[0.729, 0.731]
Sawyer	7	47	[0.798, 0.801]	0.769	[0.774, 0.777]	0.900	[0.904, 0.907]
	8	41	[0.727, 0.731]	0.679	[0.685, 0.688]	0.807	[0.811, 0.813]
	9	37	[0.764, 0.767]	0.722	[0.761, 0.763]	0.725	[0.731, 0.733]
	10	33	[0.541, 0.544]	0.517	[0.541, 0.543]	0.653	[0.657, 0.659]
	11	31	[0.318, 0.320]	0.288	[0.291, 0.292]	0.296	[0.300, 0.302]
Average			12.38	11.77	11.97	18.91	19.11

Table 6.5 Comparison of the average utility work in each station for different rules applied to instance 'Hahn' with 6 stations

Rule	Station						Total
	1	2	3	4	5	6	
FiFo	0.0	9.10	22.73	6.60	29.76	64.49	132.68
min UW	0.0	8.15	11.42	6.30	12.88	25.80	64.55
Local search	0.0	9.11	10.22	6.28	11.26	21.52	58.39

in which the variables are the utility work caused from a product p, which is expressed as UW_{ps} for station s. The differences from the 'min UW' rule are the total disregard of station 2, the decrease of importance of station 4, and some fine adjustments on stations 1 and 6. The average amount of utility work in each station for the different rules is shown in Table 6.5. Comparing the 'FiFo' and 'min UW' rules, the amount of utility work is reduced in every station (besides 1, which does not require utility work). The decrease is, however, very unevenly distributed among the stations. The reductions on stations 2 and 4, for instance, are way smaller than the reductions on stations 3, 5, and 6. One explanation for the improvement of the local search is that the weight on station 2 is completely removed and the exponent decreased for station 4. The result is that the utility work on station 2 is larger than the 'min UW' rule but considerably lower on stations 3, 5, and 6. The resulting effect is a significant reduction in the total utility work required.

The two rightmost columns of Table 6.4 refer to the local search with a random start solution. From the 45 tested instances, the random start only produced three rules that are better than the local search based on the 'min UW' rule. On average, the random start produced much poorer policies (average of 19.11 as compared to 18.31 for the 'min UW' rule).

From the available results, the local search is able to provide slightly better results than the 'min UW' rule. When starting with the values of 1 in each coefficient and exponent ('min UW' rule), the algorithm is able in most instances to perform a fine tuning of the policy. The results are, however, only marginally better. The improvement procedure can also be used with a random start rule. This approach may require multiple runs since the average results are worse than the simple rule. Other difficulties with this method relate to the identical rules that can be achieved with different expressions (by factoring all constants by 0.5, for instance). Further work should consider normalizing the policy-expression and use methods that can escape local optima.

6.5.5   The Lookahead Procedure

The second approach to improve the use of the buffer is to consider the combination of multiple products in the buffer. The lookahead procedure partially enumerates the possible sequences before selecting a model to forward to the assembly line.

In the lookahead algorithm, the required utility work of subsequences of a given length is calculated. For this, the due date of the products is considered. That is, if a product has a critical due date, it must be considered in the subsequences. The selection of a subsequence is performed based on the minimal aggregated utility work required. The first model of the best sequence is selected and sent to production. In the next iteration, a new product enters the buffer and the utility work is recalculated. If more than one sequence exhibits the same amount of utility work, the product with the lowest due date between the first models in the sequences is selected.

For this chapter, the enumeration of subsequences of 2 and 3 products are tested. The confidence interval of simulations with 500 repetitions and 50,000 products are given in Table 6.6. The 'min UW' rule can be seen as a lookahead procedure of depth one and is also listed in the table for comparison. On average, considering multiple products in the selection can significantly reduce the necessary utility work from 12.38 ('min UW') to 11.20 (lookahead with depth 2) and 10.12 (lookahead with depth 3). The explanation for the improvement is that a depth larger than 1 can consider the due date of products which will be due in the next iterations. Suppose that in iteration h there is a product in the buffer with due date $h + 1$. The 'min UW' rule would select the model with the minimal amount of utility work, even if the selection does not match well the processing times of the product with due date at $h + 1$. Using a lookahead search with depth 2, this effect is considered: the selected product is the one that in combination with the critical model in the next iteration would require the minimal amount of utility work. The results in Table 6.6 indicate that the amount of utility work decreases, on average, with the depth of the search.

However, a larger depth does not necessarily mean that the policy is better. For the instances 'Hahn' with 6, 7, and 10 stations, considering a sequence of 2 products provides worse results than the 'min UW' rule. Specifically, for instance 'Hahn' with 7 stations, the 'min UW' rule provides better results than the lookahead with 2 or 3 products. Other instances, such as 'Buxey' with 9 stations and 'Roszieg' with 8 stations prove to be almost unaffected by the larger lookahead depths. By exploring the result files of these instances, no explanation for the better performance of the simple rule rather than random effects.

Table 6.6 Comparison of the 95%-confidence intervals for the lookahead policy

Instance	NS	CT	I ('min UW')	Lookahead depth	
				2	3
Buxey	7	48	[0.419, 0.421]	[0.396, 0.398]	[0.395, 0.397]
	8	42	[0.492, 0.494]	[0.397, 0.399]	[0.382, 0.384]
	9	38	[0.213, 0.215]	[0.213, 0.214]	[0.213, 0.215]
	10	34	[0.462, 0.464]	[0.436, 0.438]	[0.435, 0.436]
	11	31	[0.478, 0.480]	[0.464, 0.467]	[0.436, 0.438]
	12	29	[0.086, 0.087]	[0.055, 0.055]	[0.053, 0.053]
Gunther	6	77	[4.85, 4.87]	[4.20, 4.22]	[3.86, 3.88]
	7	64	[3.98, 4.00]	[3.17, 3.19]	[2.82, 2.83]
	8	56	[4.70, 4.72]	[3.88, 3.90]	[3.50, 3.52]
	9	54	[1.32, 1.33]	[1.04, 1.05]	[0.971, 0.976]
Hahn	3	4659	[44.77, 45.04]	[30.31, 30.58]	[25.57, 25.79]
	4	3566	[49.02, 49.20]	[40.11, 40.29]	[37.39, 37.53]
	5	2744	[60.85, 61.04]	[53.91, 54.11]	[50.20, 50.39]
	6	2341	[64.44, 64.71]	[68.41, 68.74]	[55.92, 56.18]
	7	2123	[26.01, 26.27]	[33.02, 33.32]	[26.99, 27.23]
	8	1827	[43.95, 44.11]	[36.49, 36.63]	[33.52, 33.65]
	9	1665	[16.95, 17.05]	[13.78, 13.85]	[13.67, 13.74]
	10	1588	[7.26, 7.34]	[8.82, 8.90]	[6.13, 6.19]

(continued)

Table 6.6 (continued)

Instance	NS	CT	Lookahead depth		
			1 (′min UW′)	2	3
Heskia	3	327	[6.20, 6.24]	[5.76, 5.80]	[5.33, 5.37]
	4	246	[7.67, 7.71]	[7.26, 7.31]	[6.83, 6.87]
	5	197	[8.37, 8.40]	[7.88, 7.92]	[7.35, 7.39]
	6	164	[8.92, 8.96]	[8.51, 8.56]	[8.03, 8.08]
	7	141	[10.24, 10.28]	[9.59, 9.63]	[9.18, 9.22]
	8	123	[10.20, 10.24]	[10.08, 10.13]	[9.95, 9.99]
Lutz1	8	1743	[71.37, 71.60]	[63.86, 64.12]	[58.98, 59.18]
	9	1595	[47.51, 47.68]	[40.38, 40.52]	[38.04, 38.18]
	10	1464	[46.80, 46.95]	[42.89, 43.05]	[40.95, 41.07]
Mitchell	3	36	[0.414, 0.416]	[0.389, 0.391]	[0.387, 0.389]
	4	27	[0.664, 0.667]	[0.618, 0.620]	[0.615, 0.617]
	5	22	[0.527, 0.529]	[0.479, 0.481]	[0.475, 0.477]
	6	19	[0.271, 0.272]	[0.259, 0.260]	[0.258, 0.260]
	7	16	[0.268, 0.269]	[0.206, 0.207]	[0.198, 0.199]
	8	15	[0.229, 0.230]	[0.209, 0.211]	[0.205, 0.206]
Roszieg	4	29	[0.537, 0.539]	[0.481, 0.482]	[0.473, 0.475]
	5	23	[0.447, 0.449]	[0.331, 0.333]	[0.315, 0.316]
	6	20	[0.514, 0.516]	[0.493, 0.495]	[0.492, 0.493]
	7	18	[0.117, 0.118]	[0.112, 0.113]	[0.111, 0.112]
	8	16	[0.024, 0.025]	[0.024, 0.024]	[0.024, 0.024]
	9	13	[0.675, 0.678]	[0.605, 0.607]	[0.596, 0.598]
	10	12	[0.538, 0.540]	[0.467, 0.470]	[0.432, 0.433]
Sawyer	7	47	[0.798, 0.801]	[0.710, 0.713]	[0.699, 0.701]
	8	41	[0.727, 0.731]	[0.569, 0.572]	[0.542, 0.544]
	9	37	[0.764, 0.767]	[0.661, 0.664]	[0.657, 0.660]
	10	33	[0.541, 0.544]	[0.448, 0.451]	[0.437, 0.439]
	11	31	[0.318, 0.320]	[0.277, 0.278]	[0.270, 0.272]
Average			12.38	11.20	10.12

6.5.6   Summary of the Results

From all the reported results, the best of the proposed policies to operate a buffer at the entrance of the assembly line with random input is to use a lookahead search. On average, longer search depths produce better results. At the instances level, however, the improvement is not equally observable for all instances. 'Buxey' with 9 stations, for instance, shows almost identical results for all tested rules.

As far as the proposed methods are concerned, a depth search (limited by the available decision time) is the most efficient decision policy. Since the decisions must be performed at every cycle time, searches in large buffers can only be partial.

Conclusion 7

7.1 Summary, Objectives, and Conclusions

This manuscript presents contributions to the literature on the assembly-line balancing problem including demand uncertainty which is present, for example, in the automotive industry. The vehicle production is highly customizable and the demand relies directly on customer orders. Although the dynamic behavior of the demand is common in practice, the topic is not well represented in the assembly-line balancing literature. The main objectives of this thesis are evaluated here, along with the most important results described in each chapter.

Objective 1: description of the production process at automotive manufacturers.
The description of the several production stages and their usual organization is described in Chapter 2. The justification of the use of assembly lines, among the definition of related optimization problems, is included in the chapter. A special focus is given to the effects of the presence of multiple products in a single assembly system, as well as the influence of the production sequence. The chapter also includes the definition of paced and unpaced systems and it is briefly explained how buffers work in the automotive production system.

Objective 2: description and classification of the literature on uncertainties in assembly-line balancing.
Chapter 3 contains a vast literature review on research papers dealing with assembly-line balancing involving some sort of uncertainty. The chapter starts with the review of the existing surveys on assembly-line balancing and related problems, describing how the uncertainty is classified. The classification scheme of Boysen et al. [2007] is extended to consider the differences of how uncertainty is modeled. Contributions to assembly line balancing, sequencing, rebalancing, resequencing, disassembly, and

buffer allocation are discussed. It is noted that the uncertainty of processing times dominates the literature so that the gap of uncertain demand is shown and described.

Objective 3: definition of the problem and development of an exact solution method for the integrated assembly-line-balancing and sequencing problem under stochastic demand.

In Chapter 4, the algorithm proposed in Sikora [2021] is described and improved. The approach considers that the assembly-line planner has total control over the sequencing of the products so that the assembly-line balancing and the sequencing problem can be solved as a combined problem. As both problems have a different time frame, the approach considers a two-stage decision process. On the first stage, the assignment of the tasks is decided, before the line is built and the demand information is known. After the implementation of the line, the demand realizations occur, so that the sequencing problem deals with real customer orders. The demand is considered stochastic and is modeled based on a discrete set of scenarios. The resulting problem consists of a two-stage stochastic programming problem, in which the balancing and sequencing problems are the first and second stages. A version of the Benders' decomposition algorithm is developed to solve the problem by exploring its block structure. As the second stage contains a mixed-integer problem, combinatorial cuts are developed for the algorithm. The combinatorial cuts are mostly weak and may only cut off the tested node so that a relaxed combinatorial cut is also proposed for the algorithm. These cuts are named partial combinatorial cuts since only parts of the subproblem are solved to generate the cuts. The strength of these cuts is explored in Section 4.3.2, in which a complete enumeration of all possible cuts is performed on sample instances to estimate the provided integrality gap. Further variable reductions and valid inequalities are introduced to reduce the integrality gap of the procedure. As a complement to the method proposed in Sikora [2021], a local search routine is integrated into the Benders' decomposition framework, so that first-stage solutions are found faster. The results show a slight improvement of the algorithm comparing to the one reported in Sikora [2021]. A dataset containing 80 instances of medium size (50 tasks) is solved in a reasonable time.

Objective 4: definition of the problem and development of an exact solution method for the assembly line balancing under random production sequences.

The fourth objective is the main focus of Chapter 5. For this problem, the boundary condition is set to no control over the sequence. In the view of the assembly line planner, the product order is random, given the probabilities for each product variation. The product models are defined based on a combination of task options: discrete choices the customer has for each element. Examples are engine size and sunroof, which can be manual, electrical, or non-existent. Based on the combination of tasks with multiple options, the number of possible products grows exponentially in the number of tasks. In one of the instances solved in Chapter 5 ('Hahn'), for instance, $1.59 \cdot 10^{12}$ products are possible. In order to model and solve such a large number of variations, it is assumed that the tasks are independent from each other, as well as independence between the multiple stations. The stations are isolated by employing utility workers, who guarantee that the work is finished within the station's bounds. The aim of the optimization is to find the best task assignments by minimizing a cost function based on the line length and expected utility work. For a given station length, a Markov chain is used to calculate the expected utility work for a given assignment. This way, the station length is optimized for each assignment using a one-variable optimization algorithm. The search process is implemented within a Branch-and-Bound algorithm. The algorithm is capable to exactly solve instances of up to 53 tasks within a newly proposed dataset.

Objective 5: definition of the problem and development of heuristic policies for the operation of buffers in assembly systems for random product inputs.
In Chapter 6, a view of the problem with industrial restrictions is considered. In this problem, an assembly-line structure is given, while the operational decision of a buffer before the assembly line is the focus of the optimization, which is used to store the products, allowing some resequencing possibilities. The input to the buffer is assumed to be random. The product sequence is selected aiming at the minimization of the utility work while due dates must be met. As the decision must be made within every cycle time and the product entry order is unknown, the defined problem requires an online optimization approach. Simple heuristic policies are proposed and tested, along with a heuristic local-search improvement of the score expression, and a lookahead search procedure. For the utility work calculation, a simulation model is used. The results show that the use of a buffer on the entry of the assembly line can considerably reduce the amount of utility work needed, comparing to a random sequence.

7.2   Limitations and Future Works

One limitation of the solution method proposed in this thesis is that they are only applicable to paced assembly lines operating with utility workers. This restriction excludes other configurations, such as synchronous and asynchronous assembly lines, as well as other remedial actions such as stopping the line or correcting incompletions at the end of the line. The choice of a paced system as a focus of the algorithms is not arbitrary. In fact, most of the problem's decomposability is only possible due to the independence-between-stations assumption, which relies on the utility workers. Some algorithms such as the Benders' and the Dantzig-Wolfe decompositions, as well as a Lagrangean-relaxation approach are also developed by the author for unpaced assembly lines. The results, however, are limited to extremely small instances since all stations are interconnected by their completion times in unpaced systems.

The exact procedures of Chapters 4 and 5 also present limitations on the instance size. Even though medium-sized instances are solved in a reasonable time, real-world assembly lines may contain hundreds of tasks, which may not be solvable with the proposed algorithms using the given hardware. In the next paragraphs, each contribution chapter is described individually, showing both, their limitations and their opportunities for further work.

One important critique of an anonymous reviewer of the article Sikora [2021] is about scenario generation. In the dataset of Chapter 4 and part of the datasets of Sikora [2021], demand scenarios are considered given and are used only to test the algorithm and its components. Further work, however, is lacking in order to produce representative demand scenarios as input parameters. As discussed in Sikora [2021], the modeled demand can come from multiple sources of uncertainty. One possibility is to consider the relative demand of the products as constant while attributing the uncertainty to the random (in time) order by the customers. In this short-term scenario, the demand variation can be seen on a daily or weekly basis and is caused by ordering fluctuations. No new products are introduced or discontinued. A set of scenarios would need to represent the possible short-term fluctuations, which would be hedged by solving the balancing in the stochastic framework. These scenarios are expected to repeatedly occur during the life of the assembly line. This way, an average solution may be meaningful.

Another possibility of demand forecast is a long-term prediction. In this case, the scenarios would model the entrance, acceptance, or discontinuation of products. Although the algorithm works the same for any given set of demand scenarios, the solution of a long-term planning has a different meaning in terms of what is being hedged. Forecasting product acceptance and consumer trends with discrete scenarios may mean that only one (or even none of them) would be realized. For this case, the proposed algorithm would calculate the best assignment for the average of the scenarios, which may not be relevant. In this regard, a regret function or robust optimization may be more appropriate. Finally, the prediction models used to forecast the demand are also left as future work.

The project presented in Chapter 5 is able to optimize assembly lines containing more than 10^{12} possible products. This order of magnitude is only possible due to strong assumptions on the independence of the tasks and stations. If the sequences of two or more stations are interrelated, the station-wise decomposition is not possible, heavily reducing the size of instances which can be solved. Furthermore, the independence of tasks may not always be achievable. In practice, a vehicle with 2 or 4 doors may present different processing times for mounting the back seats, for instance. A second strong limitation on the size of the instances is that the length of the stations is a variable to be minimized. Although this decision shows that considering the length can produce better assembly-line solutions, a known lower bound for the cost is lacking. In the implementation of Chapter 5, the lower bounds of the nodes are generated dynamically, by iterating values of station length. Although a large part of the nodes is cut off before the complete computation of the optimal length, all feasible nodes must be generated, as the initial lower bound is very weak. Because of this, larger instances are limited by the memory of the computer in the generation of the search tree already. Directions for further work may consider the development of lower bounds, different enumeration schemes, or an approximative enumeration procedure for a heuristic solution.

The problem treated in Chapter 6 is an online optimization problem under uncertainty. Due to the complexity and the limitation on the decision time, only simple heuristic policies based on score expressions and an incomplete enumeration are tested. As further work, more elaborated expression-based policies can be explored considering more information than only the caused utility work. Another idea is to combine the lookahead search with the expression-based policies, instead of minimizing the resulting utility work time only.

A final critique is on the focus on traditional assembly lines using conveyor belts for the movement. A new trend in practice and the literature considers assembly lines without the restriction of a conveyor belt. In these approaches, the products are transported by Automated Guided Vehicles (AGVs), changing the flow shop to job shop hybrid systems. The flexibility in the movement can improve the task allocation since the layout needs not to be serial anymore. Furthermore, the movement of the AGVs works as a buffer, compensating in travel times or waiting times the model imbalances without stopping the whole production system. Such a concept is the focus of Hottenrott and Grunow [2019], who optimize the line layout minimizing cost and movement.

Bibliography

Ağpak, K. and Gökçen, H. [2007]. A chance-constrained approach to stochastic line balancing problem, *European Journal of Operational Research* **180**(3): 1098–1115.

Akpinar, S., Elmi, A. and Bektaş, T. [2017]. Combinatorial Benders cuts for assembly line balancing problems with setups, *European Journal of Operational Research* **259**(2): 527–537.

Al-E-Hashem, S. M., Aryanezhad, M. B., Malekly, H. and Sadjadi, S. J. [2009]. Mixed model assembly line balancing problem under uncertainty, *2009 International Conference on Computers and Industrial Engineering, CIE 2009* pp. 233–238.

Altekin, F. T. and Akkan, C. [2012]. Task-failure-driven rebalancing of disassembly lines, *International Journal of Production Research* **50**(18): 4955–4976.

Arnold, D., Isermann, H., Kuhn, A., Tempelmeier, H. and Furmans, K. [2008]. *Handbuch Logistik*, VDI-Buch, Springer Berlin Heidelberg.

Bagher, M., Zandieh, M. and Farsijani, H. [2011]. Balancing of stochastic U-type assembly lines: an imperialist competitive algorithm, *The International Journal of Advanced Manufacturing Technology* **54**(1): 271–285.

Bard, J. F. [1989]. Assembly line balancing with parallel workstations and dead time, *International Journal of Production Research* **27**(6): 1005–1018.

Bartholdi, J. J. [1993]. Balancing two-sided assembly lines: a case study, *International Journal of Production Research* **31**(10): 2447–2461.

Battaïa, O. and Dolgui, A. [2013]. A taxonomy of line balancing problems and their solution approaches, *International Journal of Production Economics* **142**(2): 259–277.

Bautista, J., Batalla-García, C. and Alfaro-Pozo, R. [2016]. Models for assembly line balancing by temporal, spatial and ergonomic risk attributes, *European Journal of Operational Research* **251**(3): 814–829.

Baybars, I. [1986]. Survey of exact algorithms for the simple assembly line balancing problem, *Management Science* **32**(8): 909–932.

Baykasoğlu, A. and Özbakır, L. [2007]. Stochastic U-line balancing using genetic algorithms, *International Journal of Advanced Manufacturing Technology* **32**(1–2): 139–147.

Becker, C. and Scholl, A. [2006]. A survey on problems and methods in generalized assembly line balancing, *European Journal of Operational Research* **168**(3): 694–715.

Becker, C. and Scholl, A. [2009]. Balancing assembly lines with variable parallel workplaces: Problem definition and effective solution procedure, *European Journal of Operational Research* **199**(2): 359–374.

Benders, J. F. [1962]. Partitioning Procedures for Solving Mixed-variables Programming Problems, *Numer. Math.* **4**(1): 238–252.

Bentaha, M. L., Dolgui, A. and Battaïa, O. [2015]. A bibliographic review of production line design and balancing under uncertainty, *IFAC-PapersOnLine* **28**(3): 70–75.

Betts, J. and Mahmoud, K. I. [1989]. Identifying multiple solutions for assembly line balancing having stochastic task times, *Computers & Industrial Engineering* **16**(3): 427–445.

Beyer, H. G. and Sendhoff, B. [2007]. Robust optimization—A comprehensive survey, *Computer Methods in Applied Mechanics and Engineering* **196**(33–34): 3190–3218.

Binder, A. K. and Rae, J. B. [2020]. Automotive industry. **URL:** https://www.britannica.com/technology/automotive-industry *(last access: 01/18/2021)*

Birge, J. R. and Louveaux, F. [2011]. *Introduction to Stochastic Programming*, 2 edn, Springer-Verlag, New York.

Boysen, N., Emde, S., Hoeck, M. and Kauderer, M. [2015]. Part logistics in the automotive industry: Decision problems, literature review and research agenda, *European Journal of Operational Research* **242**(1): 107–120.

Boysen, N. and Fliedner, M. [2008]. A versatile algorithm for assembly line balancing, *European Journal of Operational Research* **184**(1): 39–56.

Boysen, N., Fliedner, M. and Scholl, A. [2007]. A classification of assembly line balancing problems, *European Journal of Operational Research* **183**(2): 674–693.

Boysen, N., Fliedner, M. and Scholl, A. [2008]. Assembly line balancing: Which model to use when?, *International Journal of Production Economics* **111**(2): 509–528.

Boysen, N., Fliedner, M. and Scholl, A. [2009a]. Assembly line balancing: Joint precedence graphs under high product variety, *IIE Transactions* **41**(3): 183–193.

Boysen, N., Fliedner, M. and Scholl, A. [2009b]. Production planning of mixed-model assembly lines: overview and extensions, *Production Planning & Control* **20**(5): 455–471.

Boysen, N., Fliedner, M. and Scholl, A. [2009c]. Sequencing mixed-model assembly lines: Survey, classification and model critique, *European Journal of Operational Research* **192**(2): 349–373.

Boysen, N., Golle, U. and Rothlauf, F. [2011]. The Car Resequencing Problem with Pull-Off Tables, *Business Research* **4**(2): 276–292.

Boysen, N., Scholl, A. and Wopperer, N. [2012]. Resequencing of mixed-model assembly lines: Survey and research agenda, *European Journal of Operational Research* **216**(3): 594–604.

Bryton, B. [1954]. *Assembly-Line Balancing Problem*, PhD thesis, Northwestern University.

Bukchin, J., Dar-El, E. M. and Rubinovitz, J. [2002]. Mixed model assembly line design in a make-to-order environment, *Computers & Industrial Engineering* **41**(4): 405–421.

Buzacott, J. A. [1999]. The Structure of Manufacturing Systems: Insights on the Impact of Variability, *International Journal of Flexible Manufacturing Systems* **11**(2): 127–146.

Cakir, B., Altiparmak, F. and Dengiz, B. [2011]. Multi-objective optimization of a stochastic assembly line balancing: A hybrid simulated annealing algorithm, *Computers and Industrial Engineering* **60**(3): 376–384.

Carraway, R. L. [1989]. A Dynamic Programming Approach to Stochastic Assembly Line Balancing, *Management Science* **35**(4): 459–471.

Carter, J. C. and Silverman, F. N. [1984]. A cost-effective approach to stochastic line balancing with off-line repairs, *Journal of Operations Management* **4**(2): 145–157.

Caserta, M. and Voβ, S. [2020]. Accelerating mathematical programming techniques with the corridor method, *International Journal of Production Research* in press.

Cechin, R. B. and Corso, L. L. [2019]. High-order multivariate Markov chain applied in Dow Jones and IBOVESPA indexes, *Pesquisa Operacional* **39**(1): 205–223.

Chakravarty, A. K. [1988]. Line balancing with task learning effects, *IIE Transactions (Institute of Industrial Engineers)* **20**(2): 186–193.

Chakravarty, A. K. and Shtub, A. [1986]. A cost minimization procedure for mixed model production lines with normally distributed task times, *European Journal of Operational Research* **23**(1): 25–36.

Chiang, W.-C. C. and Urban, T. L. [2006]. The stochastic U-line balancing problem: A heuristic procedure, *European Journal of Operational Research* **175**(3): 1767–1781.

Chiang, W.-C. C., Urban, T. L. and Luo, C. [2016]. Balancing stochastic two-sided assembly lines, *International Journal of Production Research* **54**(20): 6232–6250.

Chica, M., Bautista, J., Cordón, Ó. and Damas, S. [2016]. A multiobjective model and evolutionary algorithms for robust time and space assembly line balancing under uncertain demand, *Omega* **58**: 55–68.

Chica, M., Bautista, J. and de Armas, J. [2019]. Benefits of robust multiobjective optimization for flexible automotive assembly line balancing, *Flexible Services and Manufacturing Journal* **31**(1): 75–103.

Chica, M., Cordón, Ó., Damas, S. and Bautista, J. [2013]. A robustness information and visualization model for time and space assembly line balancing under uncertain demand, *International Journal of Production Economics* **145**(2): 761–772.

Chutima, P. and Yiangkamolsing, C. [2003]. Application of fuzzy genetic algorithm for sequencing in mixed-model assembly line with processing time, *The International Journal of Industrial Engineering: Theory, Applications and Practice* **10**(4): 325–331.

Codato, G. and Fischetti, M. [2006]. Combinatorial Benders' Cuts for Mixed-Integer Linear Programming, *Operations Research* **54**(4): 756–766.

Costa, A. M., Cordeau, J.-F., Gendron, B. and Laporte, G. [2012]. Accelerating benders decomposition with heuristic master problem solutions, *Pesquisa Operacional* **32**(1): 3–19.

Dar-El, E. M. and Nadivi, A. [1981]. A mixed-model sequencing application, *International Journal of Production Research* **19**(1): 69–84.

Delice, Y., Kizilkaya Aydoğan, E. and Özcan, U. [2016]. Stochastic two-sided U-Type assembly line balancing: A genetic algorithm approach, *International Journal of Production Research* **54**(11): 3429–3451.

Demir, L., Tunali, S. and Eliiyi, D. T. [2014]. The state of the art on buffer allocation problem: a comprehensive survey, *Journal of Intelligent Manufacturing* **25**(3): 371–392.

Dolgui, A. and Kovalev, S. [2012]. Scenario based robust line balancing: Computational complexity, *Discrete Applied Mathematics* **160**(13–14): 1955–1963.

Dong, J., Zhang, L. and Xiao, T. [2018]. A hybrid PSO/SA algorithm for bi-criteria stochastic line balancing with flexible task times and zoning constraints, *Journal of Intelligent Manufacturing* **29**(4): 737–751.

Dong, J., Zhang, L., Xiao, T. and Mao, H. [2014]. Balancing and sequencing of stochastic mixed-model assembly U-lines to minimise the expectation of work overload time, *International Journal of Production Research* **52**(24): 7529–7548.

Drexl, A. and Kimms, A. [2001]. Sequencing JIT Mixed-Model Assembly Lines Under Station-Load and Part-Usage Constraints, *Management Science* **47**(3): 480–491.

Drexl, A., Kimms, A. and Matthießen, L. [2006]. Algorithms for the car sequencing and the level scheduling problem, *Journal of Scheduling* **9**(2): 153–176.

Eghtesadifard, M., Khalifeh, M. and Khorram, M. [2020]. A systematic review of research themes and hot topics in assembly line balancing through the web of science within 1990–2017, *Computers & Industrial Engineering* **139**: 106182.

Emde, S., Boysen, N. and Scholl, A. [2010]. Balancing mixed-model assembly lines: A computational evaluation of objectives to smoothen workload, *International Journal of Production Research* **48**(11): 3173–3191.

Emde, S. and Gendreau, M. [2017]. Scheduling in-house transport vehicles to feed parts to automotive assembly lines, *European Journal of Operational Research* **260**(1): 255–267.

Erel, E., Sabuncuoglu, I. and Sekerci, H. [2005]. Stochastic assembly line balancing using beam search, *International Journal of Production Research* **43**(7): 1411–1426.

Erel, E. and Sarin, S. C. [1998]. A survey of the assembly line balancing procedures.

European Automobile Manufacturers Association (ACEA) [2019]. Vehicles in use—Europe 2019. **URL:** https://www.acea.be/publications/article/report-vehicles-in-use-europe-2019 *(last access: 01/18/2021)*

European Automobile Manufacturers Association (ACEA) [2020]. The Automobile Industry Pocket Guide 2020–2021. **URL:** https://www.acea.be/uploads/publications/ACEA_Pocket_Guide_2020-2021.pdf *(last access: 01/18/2021)*

Falkenauer, E. [2005]. Line balancing in the real world, *Proceedings of the International Conference on Product Lifecycle Management, PLM 05*, Lyon, pp. 1–10.

Fattahi, P. and Salehi, M. [2009]. Sequencing the mixed-model assembly line to minimize the total utility and idle costs with variable launching interval, *The International Journal of Advanced Manufacturing Technology* **45**(9): 987.

Ferguson, D. E. [1960]. Fibonaccian Searching, *Commununications of the ACM* **3**(12): 648. **URL:** https://doi.org/10.1145/367487.367496

Fischetti, M., Ljubić, I. and Sinnl, M. [2016]. Benders decomposition without separability: A computational study for capacitated facility location problems, *European Journal of Operational Research* **253**(3): 557–569.

Fischetti, M., Ljubić, I. and Sinnl, M. [2017]. Redesigning Benders Decomposition for Large-Scale Facility Location, *Management Science* **63**(7): 2146–2162.

Fleszar, K. and Hindi, K. S. [2003]. An enumerative heuristic and reduction methods for the assembly line balancing problem, *European Journal of Operational Research* **145**(3): 606–620.

Fliedner, M. and Boysen, N. [2008]. Solving the car sequencing problem via Branch & Bound, *European Journal of Operational Research* **191**(3): 1023–1042.

Gagné, C., Gravel, M. and Price, W. L. [2006]. Solving real car sequencing problems with ant colony optimization, *European Journal of Operational Research* **174**(3): 1427–1448.

Gagnon, R. and Ghosh, S. [1991]. Assembly line research: Historical roots, research life cycles and future directions, *Omega* **19**(5): 381–399.

Gamberini, R., Gebennini, E., Grassi, A. and Regattieri, A. [2009]. A multiple single-pass heuristic algorithm solving the stochastic assembly line rebalancing problem, *International Journal of Production Research* **47**(8): 2141–2164.

Gamberini, R., Grassi, A. and Rimini, B. [2006]. A new multi-objective heuristic algorithm for solving the stochastic assembly line re-balancing problem, *International Journal of Production Economics* **102**(2): 226–243.

Gershwin, S. B. and Schor, J. E. [2000]. Efficient algorithms for buffer space allocation, *Annals of Operations Research* **93**(1): 117–144.

Ghosh, S. and Gagnon, R. J. [1989]. Comprehensive literature review and analysis of the design, balancing and scheduling of assembly systems, *International Journal of Production Research* **27**(4): 637–670.

Golle, U., Rothlauf, F. and Boysen, N. [2014a]. Car sequencing versus mixed-model sequencing: A computational study, *European Journal of Operational Research* **237**(1): 50–61.

Golle, U., Rothlauf, F. and Boysen, N. [2014b]. Iterative beam search for car sequencing, *Annals of Operations Research* **226**(1): 239–254.

Guerriero, F. and Miltenburg, J. [2003]. The stochastic U-line balancing problem, *Naval Research Logistics* **50**(1): 31–57.

Gungor, A. and Gupta, S. M. [2001]. A solution approach to the disassembly line balancing problem in the presence of task failures, *International Journal of Production Research* **39**(7): 1427–1467.

Günther, H. O. and Tempelmeier, H. [2012]. *Produktion und Logistik*, Springer-Lehrbuch, 9 edn, Springer Berlin Heidelberg.

Gurevsky, E., Battaïa, O. and Dolgui, A. [2012]. Balancing of simple assembly lines under variations of task processing times, *Annals of Operations Research* **201**(1): 265–286.

Gurevsky, E., Battaïa, O. and Dolgui, A. [2013]. Stability measure for a generalized assembly line balancing problem, *Discrete Applied Mathematics* **161**(3): 377–394.

Gurevsky, E., Hazir, Ö., Battaïa, O. and Dolgui, A. [2013]. Robust balancing of straight assembly lines with interval task times, *Journal of the Operational Research Society* **64**(11): 1607–1613.

Gutjahr, A. and Nemhauser, G. [1964]. An algorithm for the line balancing problem, *Management Science* **11**(2): 308–315.

Gwiggner, C. [2020]. Personal communication: A priori sequencing with uncertain release dates.

Hamidinejad, S. M., Kolahan, F. and Kokabi, A. H. [2012]. The modeling and process analysis of resistance spot welding on galvanized steel sheets used in car body manufacturing, *Materials and Design* **34**: 759–767.

Han, M.-S. and Park, D.-J. [2002]. Optimal buffer allocation of serial production lines with quality inspection machines, *Computers & Industrial Engineering* **42**(1): 75–89.

Hazir, Ö., Dolgui, A., Hazır, Ö. and Dolgui, A. [2013]. Assembly line balancing under uncertainty: Robust optimization models and exact solution method, *Computers & Industrial Engineering* **65**(2): 261–267.

Hazir, Ö., Dolgui, A., Hazır, Ö., Dolgui, A., Hazır, Ö. and Dolgui, A. [2015]. A decomposition based solution algorithm for U-type assembly line balancing with interval data, *Computers & Operations Research* **59**: 126–131.

Henig, M. I. [1986]. Extensions of the dynamic programming method in the deterministic and stochastic assembly-line balancing problems, *Computers and Operations Research* **13**(4): 443–449.

Hillier, F. S. and Boling, R. W. [1979]. On the Optimal Allocation of Work in Symmetrically Unbalanced Production Line Systems with Variable Operation Times, *Management Science* **25**(8): 721–728.

Hillier, M. [2013]. Designing unpaced production lines to optimize throughput and work-in-process inventory, *IIE Transactions* **45**(5): 516–527.

Hillier, M. S. and Hillier, F. S. [2006]. Simultaneous optimization of work and buffer space in unpaced production lines with random processing times, *IIE Transactions* **38**(1): 39–51.

Hoffmann, T. [1963]. Assembly line balancing with a precedence matrix, *Management Science* **9**(4): 551–562.

Hop, N. V. [2006]. A heuristic solution for fuzzy mixed-model line balancing problem, *European Journal of Operational Research* **168**(3): 798–810.

Hottenrott, A. and Grunow, M. [2019]. Flexible layouts for the mixed-model assembly of heterogeneous vehicles, *OR Spectrum* **41**(4): 943–979.

Hudson, S., McNamara, T. and Shaaban, S. [2015]. Unbalanced lines: where are we now?, *International Journal of Production Research* **53**(6): 1895–1911.

International Organization of Motor Vehicle Manufacturers (OICA) [2017]. World Motor Vehicle Production OICA correspondents survey—World Ranking of Manufacturers. **URL:** http://www.oica.net/category/production-statistics/2017-statistics/ *(last access: 01/18/2021)*

International Organization of Motor Vehicle Manufacturers (OICA) [2018]. World motor vehicle production by country and type. **URL:** http://www.oica.net/category/production-statistics/2018-statistics/ *(last access: 01/18/2021)*

Jackson, J. [1956]. A computing procedure for a line balancing problem, *Management Science* **2**(3): 261–271.

Kampker, A., Burggräf, P., Kreisköther, K. D., Dannapfel, M., Bertram, S. and Wagner, J. [2017]. Flexibility through mobility: the e-mobile assembly of tomorrow, *7. WGP-Jahreskongress Aachen*, Apprimus Verlag, Aachen, pp. 269–294.

Kao, E. P. [1979]. Computational experience with a stochastic assembly line balancing algorithm, *Computers & Operations Research* **6**(2): 79–86.

Kao, E. P. C. [1976]. A Preference Order Dynamic Program for Stochastic Assembly Line Balancing, *Management Science* **22**(10): 1097–1104.

Karabati, S. and Sayın, S. [2003]. Assembly line balancing in a mixed-model sequencing environment with synchronous transfers, *European Journal of Operational Research* **149**(2): 417–429.

Kiefer, J. [1953]. Sequential minimax search for a maximum, *Proceedings of the American Mathematical Society* **4**: 502–506.

Kottas, J. F. and Lau, H. S. [1973]. A cost-oriented approach to stochastic line balancing, *AIIE Transactions* **5**(2): 164–171.

Kottas, J. F. and Lau, H. S. [1976]. A total operating cost model for paced lines with stochastic task times, *AIIE Transactions* **8**(2): 234–240.

Kottas, J. and Lau, H. [1981]. A stochastic line balancing procedure, *International Journal of Production Research* **19**(2): 177–193.

Lai, T. C., Sotskov, Y. N. and Dolgui, A. [2019]. The stability radius of an optimal line balance with maximum efficiency for a simple assembly line, *European Journal of Operational Research* **274**(2): 466–481.

Lai, T. C., Sotskov, Y. N., Dolgui, A. and Zatsiupa, A. [2016]. Stability radii of optimal assembly line balances with a fixed workstation set, *International Journal of Production Economics* **182**: 356–371.

Laporte, G. and Louveaux, F. V. [1993]. The integer L-shaped method for stochastic integer programs with complete recourse, *Operations Research Letters* **13**(3): 133–142.

Leitold, D., Vathy-Fogarassy, A. and Abonyi, J. [2019]. Empirical working time distribution-based line balancing with integrated simulated annealing and dynamic programming, *Central European Journal of Operations Research* **27**(2): 455–473.

Li, J. and Gao, J. [2014]. Balancing manual mixed-model assembly lines using overtime work in a demand variation environment, *International Journal of Production Research* **52**(12): 3552–3567.

Liu, S. B., Ong, H. L. and Huang, H. C. [2005]. A bidirectional heuristic for stochastic assembly line balancing type II problem, *International Journal of Advanced Manufacturing Technology* **25**(1–2): 71–77.

Lopes, T. C., Michels, A. S., Sikora, C. G. S., Molina, R. G. and Magatão, L. [2018]. Balancing and cyclically sequencing synchronous, asynchronous, and hybrid unpaced assembly lines, *International Journal of Production Economics* **203**: 216–224.

Lopes, T. C., Sikora, C. G., Molina, R. G., Schibelbain, D., Rodrigues, L. C. and Magatão, L. [2017]. Balancing a robotic spot welding manufacturing line: An industrial case study, *European Journal of Operational Research* **263**(3): 1033–1048.

Lopes, T. C., Sikora, C. G. S., Michels, A. S., Lindbeck da Silva, A. C. and Magatão, L. [2018]. Modeling the Stochastic Steady-State of Mixed-Model Asynchronous Assembly Lines with Markov Chains, *Proceedings of the XIX Latin-Iberoamerican Conference on Operations Research (CLAIO)*, Lima, pp. 183–190.

Lopes, T. C., Sikora, C. G. S., Michels, A. S. and Magatão, L. [2020a]. An iterative decomposition for asynchronous mixed-model assembly lines: combining balancing, sequencing, and buffer allocation, *International Journal of Production Research* **58**(2): 615–630.

Lopes, T. C. T., Michels, A. A. S., Sikora, C. C. G. S. and Magatão, L. [2019]. Balancing and cyclical scheduling of asynchronous mixed-model assembly lines with parallel stations, *Journal of Manufacturing Systems* **50**: 193–200.

Lopes, T. C. T., Sikora, C. G. S. C. C. G. S., Michels, A. A. S. and Magatão, L. [2020b]. Mixed-model assembly lines balancing with given buffers and product sequence: model, formulation comparisons, and case study, *Annals of Operations Research* **286**(1–2): 475–500.

Lyu, J. J. [1997]. A Single-Run Optimization Algorithm for Stochastic Assembly Line Balancing Problems, *Journal of Manufacturing Systems* **16**(3): 204–210.

Manavizadeh, N., Rabbani, M., Moshtaghi, D. and Jolai, F. [2012]. Mixed-model assembly line balancing in the make-to-order and stochastic environment using multi-objective evolutionary algorithms, *Expert Systems with Applications* **39**(15): 12026–12031.

McCormick, S. T., Pinedo, M. L., Shenker, S. and Wolf, B. [1989]. Sequencing in an Assembly Line with Blocking to Minimize Cycle Time, *Operations Research* **37**(6): 925–935.

McMullen, P. R. and Frazier, G. V. [1997]. A heuristic for solving mixed-model line balancing problems with stochastic task durations and parallel stations, *International Journal of Production Economics* **51**(3): 177–190.

McMullen, P. R., Frazier, G. V., Mc mulen, P. R. and Frazier, G. V. [1998]. Using simulated annealing to solve a multiobjective assembly line balancing problem with parallel workstations, *International Journal of Production Research* **36**(10): 2717–2741.

McMullen, P. R. and Tarasewich, P. [2003]. Using Ant Techniques to Solve the Assembly Line Balancing Problem, *IIE Transactions* **35**(7): 605–617.

McMullen, P. R. and Tarasewich, P. [2006]. Multi-objective assembly line balancing via a modified ant colony optimization technique, *International Journal of Production Research* **44**(1): 27–42.

Meissner, S. [2010]. Controlling just-in-sequence flow-production, *Logistics Research* **2**(1): 45–53.

Meyr, H. [2004]. Supply chain planning in the German automotive industry, *OR Spectrum* **26**(4): 447–470.

Michalos, G., Makris, S., Papakostas, N., Mourtzis, D. and Chryssolouris, G. [2010]. Automotive assembly technologies review: challenges and outlook for a flexible and adaptive approach, *CIRP Journal of Manufacturing Science and Technology* **2**(2): 81–91.

Michels, A. S., Lopes, T. C., Sikora, C. G. S. and Magatão, L. [2018]. The Robotic Assembly Line Design (RALD) problem: Model and case studies with practical extensions, *Computers and Industrial Engineering* **120**: 320–333.

Michels, A. S., Lopes, T. C., Sikora, C. G. S. and Magatão, L. [2019]. A Benders' decomposition algorithm with combinatorial cuts for the multi-manned assembly line balancing problem, *European Journal of Operational Research* **278**(3): 796–808.

Miltenburg, J. [1989]. Level Schedules for Mixed-Model Assembly Lines in Just-in-Time Production Systems, *Management Science* **35**(2): 192–207.

Miltenburg, J. and Wijngaard, J. [1994]. U-line line balancing problem, *Management Science* **40**(10): 1378–1388.

Miralles, C., García-Sabater, J. P., Andrés, C. and Cardos, M. [2007]. Advantages of assembly lines in Sheltered Work Centres for Disabled. A case study, *International Journal of Production Economics* **110**(1–2): 187–197.

Moodie, C. L. and Young, H. H. [1965]. A heuristic method of assembly line balancing for assumptions of constant or variable work element times, *Journal of Industrial Engineering* **16**: 23–29.

Moreira, M. C. O., Cordeau, J.-F., Costa, A. M. and Laporte, G. [2015]. Robust assembly line balancing with heterogeneous workers, *Computers and Industrial Engineering* **88**: 254–263.

Müller, C., Grunewald, M. and Spengler, T. S. [2016]. Redundant Configuration of Automated Flow Lines, *IFAC-PapersOnLine* **49**(12): 751–756.

Müller, C., Grunewald, M. and Spengler, T. S. [2018]. Redundant configuration of robotic assembly lines with stochastic failures, *International Journal of Production Research* **56**(10): 3662–3682.

Nahas, N., Nourelfath, M. and Ait-Kadi, D. [2009]. Selecting machines and buffers in unreliable series-parallel production lines, *International Journal of Production Research* **47**(14): 3741–3774.

Nahas, N., Nourelfath, M. and Gendreau, M. [2014]. Selecting machines and buffers in unreliable assembly/disassembly manufacturing networks, *International Journal of Production Economics* **154**: 113–126.

Nkasu, M. M. M. and Leung, K. H. H. [1995]. A stochastic approach to assembly line balancing, *International Journal of Production Research* **33**(4): 975–991.

Nourelfath, M., Nahas, N. and Ait-Kadi, D. [2005]. Optimal design of series production lines with unreliable machines and finite buffers, *Journal of Quality in Maintenance Engineering* **11**(2): 121–138.

Nourmohammadi, A., Eskandari, H. and Fathi, M. [2019]. Design of stochastic assembly lines considering line balancing and part feeding with supermarkets, *Engineering Optimization* **51**(1): 63–83.

Omar, M. A. [2011]. *The automotive body manufacturing systems and processes*, John Wiley & Sons.

Otto, A., Otto, C. and Scholl, A. [2013]. Systematic data generation and test design for solution algorithms on the example of SALBPGen for assembly line balancing, *European Journal of Operational Research* **228**(1): 33–45.

Otto, A. and Scholl, A. [2011]. Incorporating ergonomic risks into assembly line balancing, *European Journal of Operational Research* **212**(2): 277–286.

Özcan, U. [2010]. Balancing stochastic two-sided assembly lines: A chance-constrained, piecewise-linear, mixed integer program and a simulated annealing algorithm, *European Journal of Operational Research* **205**(1): 81–97.

Özcan, U., Kellegöz, T. and Toklu, B. [2011]. A genetic algorithm for the stochastic mixed-model U-line balancing and sequencing problem, *International Journal of Production Research* **49**(6): 1605–1626.

Özceylan, E., Kalayci, C. B., Güngör, A. and Gupta, S. M. [2019]. Disassembly line balancing problem: a review of the state of the art and future directions, *International Journal of Production Research* **57**(15–16): 4805–4827.

Özceylan, E. and Paksoy, T. [2014]. Interactive fuzzy programming approaches to the strategic and tactical planning of a closed-loop supply chain under uncertainty, *International Journal of Production Research* **52**(8): 2363–2387.

Öztürk, C., Tunali, S., Hnich, B. and Örnek, A. [2015]. Cyclic scheduling of flexible mixed model assembly lines with paralel stations, *Journal of Manufacturing Systems* **36**(1): 147–158.

Paksoy, T., Güngör, A., Özceylan, E. and Hancilar, A. [2013]. Mixed model disassembly line balancing problem with fuzzy goals, *International Journal of Production Research* **51**(20): 6082–6096.

Patterson, J. H. and Albracht, J. J. [1975]. Assembly-Line Balancing: Zero-One Programming with Fibonacci Search, *Operations Research* **23**(1): 166–172.

Pereira, J. [2018]. The robust (minmax regret) assembly line worker assignment and balancing problem, *Computers & Operations Research* **93**: 27–40.

Pereira, J. and Álvarez-Miranda, E. [2018]. An exact approach for the robust assembly line balancing problem, *Omega* **78**: 85–98.

Rahmaniani, R., Crainic, T. G., Gendreau, M. and Rei, W. [2017]. The Benders decomposition algorithm: A literature review, *European Journal of Operational Research* **259**(3): 801–817.

Raouf, A. and Tsui, C. [1982]. A new method for assembly line balancing having stochastic work elements, *Computers & Industrial Engineering* **6**(2): 131–148.

Reeve, N. R. and Thomas, W. H. [1973]. Balancing stochastic assembly lines, *AIIE Transactions* **5**(3): 223–229.

Rossi, A., Gurevsky, E., Battaïa, O. and Dolgui, A. [2016]. Maximizing the robustness for simple assembly lines with fixed cycle time and limited number of workstations, *Discrete Applied Mathematics* **208**: 123–136.

Rubinovitz, J. and Bukchin, J. [1991]. Design and balancing of robotic assembly lines, *Proc. of the fourth world conference on robotics research*, Pittsburgh, PA.

Saif, U., Guan, Z., Liu, W., Zhang, C. and Wang, B. [2014]. Pareto based artificial bee colony algorithm for multi objective single model assembly line balancing with uncertain task times, *Computers & Industrial Engineering* **76**(1): 1–15.

Saif, U., Guan, Z., Wang, B. and Mirza, J. [2014]. Pareto lexicographic α-robust approach and its application in robust multi objective assembly line balancing problem, *Frontiers of Mechanical Engineering* **9**(3): 257–264.

Saif, U., Guan, Z., Zhang, L., Mirza, J. and Lei, Y. [2017]. Hybrid Pareto artificial bee colony algorithm for assembly line balancing with task time variations, *International Journal of Computer Integrated Manufacturing* **30**(2–3): 255–270.

Salveson, M. [1955]. The assembly line balancing problem, *Journal of Industrial Engineering* **6**(3): 18–25.

Sarin, S. C. and Erel, E. [1990]. Development of cost model for the single-model stochastic assembly line balancing problem, *International Journal of Production Research* **28**(7): 1305–1316.

Sarin, S. C., Erel, E. and Dar-El, E. M. [1999]. A methodology for solving single-model, stochastic assembly line balancing problem, *Omega* **27**(5): 525–535.

Scholl, A. [1993]. Data of Assembly Line Balancing Problems, *Technical report*, Darmstadt Technical University, Department of Business Administration, Economics and Law, Institute for Business Studies (BWL), Darmstadt.

Scholl, A. [1999]. *Balancing and Sequencing of Assembly Lines*, second edn, Physica, Heidelberg.

Scholl, A. and Becker, C. [2006]. State-of-the-art exact and heuristic solution procedures for simple assembly line balancing, *European Journal of Operational Research* **168**(3): 666–693.

Scholl, A., Boysen, N. and Fliedner, M. [2009]. Optimally solving the alternative subgraphs assembly line balancing problem, *Annals of Operations Research* **172**(1): 243–258.

Scholl, A. and Klein, R. [1997]. SALOME: A Bidirectional Branch-and-Bound Procedure for Assembly Line Balancing, *INFORMS Journal on Computing* **9**(4): 319–334.

Schumacher, M., Kreiskoether, K. D. and Kampker, A. [2018]. Proactive Resequencing of the Vehicle Order in Automotive Final Assembly to Minimize Utility Work, *Journal of Industrial and Intelligent Information* **6**(1): 1–5.

Sewell, E. C. and Jacobson, S. H. [2012]. A Branch, Bound, and Remember Algorithm for the Simple Assembly Line Balancing Problem, *INFORMS Journal on Computing* **24**(3): 433–442.

Shin, D. [1990]. An efficient heuristic for solving stochastic assembly line balancing problems, *Computers and Industrial Engineering* **18**(3): 285–295.

Shin, D. and Min, H. [1991]. Uniform Assembly Line Balancing with Stochastic Task Times in Just-in-time Manufacturing, *International Journal of Operations & Production Management* **11**(8): 23–34.

Shtub, A. [1984]. The effect of incompletion cost on line balancing with multiple manning of work stations, *International Journal of Production Research* **22**(2): 235–245.

Sikora, C. G. S. [2021]. Benders' decomposition for the balancing of assembly lines with stochastic demand, *European Journal of Operational Research* **292**(1): 108–124.

Sikora, C. G. S., Lopes, T. C. and Magatão, L. [2017]. Traveling worker assembly line (re)balancing problem: Model, reduction techniques, and real case studies, *European Journal of Operational Research* **259**(3): 949–971.

Sikora, C. G. S., Lopes, T. C., Schibelbain, D. and Magatão, L. [2017]. Integer Based Formulation for the Simple Assembly Line Balancing Problem with Multiple Identical Tasks, *Computers & Industrial Engineering* **104**: 134–144.

Silverman, F. N. and Carter, J. C. [1986]. Cost-Based Methodology for Stochastic Line Balancing With Intermittent Line Stoppages., *Management Science* **32**(4): 455–463.

Simaria, A. S., Zanella De Sá, M. and Vilarinho, P. M. [2009]. Meeting demand variation using flexible U-shaped assembly lines, *International Journal of Production Research* **47**(14): 3937–3955.

Smunt, T. L. and Perkins, W. C. [1985]. Stochastic unpaced line design: Review and further experimental results, *Journal of Operations Management* **5**(3): 351–373.

Sniedovich, M. [1981]. Analysis of a Preference Order Assembly Line Problem, *Management Science* **27**(9): 1067–1080.

Sotskov, Y. N., Dolgui, A. and Portmann, M.-C. [2006]. Stability analysis of an optimal balance for an assembly line with fixed cycle time, *European Journal of Operational Research* **168**(3): 783–797.

Sphicas, G. P. and Silverman, F. N. [1976]. Deterministic equivalents for stochastic assembly line balancing, *AIIE Transactions* **8**(2): 280–282.

Spieckermann, S., Gutenschwager, K. and Voß, S. [2004]. A sequential ordering problem in automotive paint shops, *International Journal of Production Research* **42**(9): 1865–1878.

Spinellis, D. D. and Papadopoulos, C. T. [2001]. Modular production line optimization: The exPLORE architecture, *Mathematical Problems in Engineering* **6**: 531895.

Spinellis, D., Papadopoulos, C. and Smith, J. M. [2000]. Large production line optimization using simulated annealing, *International Journal of Production Research* **38**(3): 509–541.

Suresh, G. and Sahu, S. [1994]. Stochastic assembly line balancing using simulated annealing, *International Journal of Production Research* **32**(8): 1801–1810.

Suresh, G., Vinod, V. V. and Sahu, S. [1996]. A genetic algorithm for assembly line balancing, *Production Planning and Control* **7**(1): 38–46.

Tang, Q., Li, Z., Zhang, L. and Zhang, C. [2017]. Balancing stochastic two-sided assembly line with multiple constraints using hybrid teaching-learning-based optimization algorithm, *Computers & Operations Research* **82**: 102–113.

Taube, F. and Minner, S. [2018]. Resequencing mixed-model assembly lines with restoration to customer orders, *Omega* **78**: 99–111.

Thomopoulos, N. T. [1967]. Line Balancing-Sequencing for Mixed-Model Assembly, *Management Science* **14**(2): 59–75.

Thomopoulos, N. T. [1970]. Mixed Model Line Balancing with Smoothed Station Assignments, *Management Science* **16**(9): 593–603.

Tiacci, L. [2015a]. Coupling a genetic algorithm approach and a discrete event simulator to design mixed-model un-paced assembly lines with parallel workstations and stochastic task times, *International Journal of Production Economics* **159**: 319–333.

Tiacci, L. [2015b]. Simultaneous balancing and buffer allocation decisions for the design of mixed-model assembly lines with parallel workstations and stochastic task times, *International Journal of Production Economics* **162**: 201–215.

Tiacci, L. and Mimmi, M. [2018]. Integrating ergonomic risks evaluation through OCRA index and balancing/sequencing decisions for mixed model stochastic asynchronous assembly lines, *Omega* **78**: 112–138.

Tsai, L.-H. [1995]. Mixed-Model Sequencing to Minimize Utility Work and the Risk of Conveyor Stoppage, *Management Science* **41**(3): 485–495.

Tsujimura, Y., Gen, M. and Kubota, E. [1995]. Solving fuzzy assembly-line balancing problem with genetic algorithms, *Computers & Industrial Engineering* **29**(1–4): 543–547.

Tuncel, E., Zeid, A. and Kamarthi, S. [2014]. Solving large scale disassembly line balancing problem with uncertainty using reinforcement learning, *Journal of Intelligent Manufacturing* **25**(4): 647–659.

Turowski, M., Morgan, M. and Tang, Y. [2005]. Disassembly line design with uncertainty, *Conference Proceedings—IEEE International Conference on Systems, Man and Cybernetics* **1**: 954–959.

Urban, T. L. and Chiang, W.-C. [2006]. An optimal piecewise-linear program for the U-line balancing problem with stochastic task times, *European Journal of Operational Research* **168**(3): 771–782.

Vrat, P. and Virani, A. [1976]. A cost model for optimal mix of balanced stochastic assembly line and the modular assembly system for a customer oriented production system, *International Journal of Production Research* **14**(4): 445–463.

Wee, T. and Magazine, M. [1982]. Assembly line balancing as generalized bin packing, *Operations Research Letters* **1**(2): 56–58.

Weiss, S., Schwarz, J. A. and Stolletz, R. [2019]. The buffer allocation problem in production lines: Formulations, solution methods, and instances, *IISE Transactions* **51**(5): 456–485.

Xu, W. and Xiao, T. [2009]. Robust balancing of mixed model assembly line, *COMPEL—The International Journal for Computation and Mathematics in Electrical and Electronic Engineering* **28**(6): 1489–1502.

Xu, W. and Xiao, T. [2011]. Strategic Robust Mixed Model Assembly Line Balancing Based on Scenario Planning, *Tsinghua Science & Technology* **16**(3): 308–314.

Yano, C. A. and Rachamadugu, R. [1991]. Sequencing to Minimize Work Overload in Assembly Lines with Product Options, *Management Science* **37**(5): 572–586.

Zacharia, P. and Nearchou, A. C. [2013]. A meta-heuristic algorithm for the fuzzy assembly line balancing type-E problem, *Computers & Operations Research* **40**(12): 3033–3044.

Zacharia, P. T. and Nearchou, A. C. [2012]. Multi-objective fuzzy assembly line balancing using genetic algorithms, *Journal of Intelligent Manufacturing* **23**(3): 615–627.

Zhang, W., Xu, W. and Gen, M. [2014]. Hybrid Multiobjective Evolutionary Algorithm for Assembly Line Balancing Problem with Stochastic Processing Time, *Procedia Computer Science* **36**(C): 587–592.

Zhang, W., Xu, W., Liu, G. and Gen, M. [2017]. An effective hybrid evolutionary algorithm for stochastic multiobjective assembly line balancing problem.

The manufacturer's authorised representative in the EU is Springer
Nature Customer Service Centre GmbH, Europaplatz 3, 69115 Heidelberg,
Germany. If you have any concerns regarding our products, please
contact ProductSafety@springernature.com

Printed and bound by CPI Group (UK) Ltd, Croydon, CR0 4YY
28/04/2026
02098468-0001